A spin- and momentum-resolved photoemission study of strong electron correlation in Co/Cu(001)

Dissertation

zur Erlangung des akademischen Grades
doctor rerum naturalium (Dr. rer. nat.)

der
Naturwissenschaftlichen Fakultät II
der Martin–Luther–Universität Halle–Wittenberg

vorgelegt von
HERRN DIPL.-PHYS. MARTIN ELLGUTH
geb. am 21.03.1984 in Eilenburg

Bibliografische Information der Deutschen Nationalbibliothek

Die Deutsche Nationalbibliothek verzeichnet diese Publikation in der
Deutschen Nationalbibliografie; detaillierte bibliografische Daten sind
im Internet über http://dnb.d-nb.de abrufbar.

ISBN 978-3-8325-4002-9

Logos Verlag Berlin GmbH
Comeniushof, Gubener Str. 47,
10243 Berlin
Tel.: +49 (0)30 42 85 10 90
Fax: +49 (0)30 42 85 10 92
INTERNET: http://www.logos-verlag.de

Gutachter:

1. Prof. Dr. Jürgen Kirschner
2. Prof. Dr. Wolf Widdra
3. Prof. Dr. Claus M. Schneider

Tag der Verteidigung: 07.Mai.2015

Contents

Chapter 1

Introduction

Cobalt is an interesting material for a variety of reasons. From the application side, it plays a role in magnetic recording media, spin valves [1, 2], magnetic multilayers with magneto-resistance [3], hard drive sensors (structures containing a layer of a cobalt alloy that show a giant-magneto-resistance effect) [4]. Furthermore, the injection of spin-polarized currents into semiconductors or into two-dimensional electron gases (spintronics) is a prospective application of ferromagnets, in general [5–7].

Cobalt is one of the elemental ferromagnets within the series of 3d transition metals. As such it exhibits a number of physical pecularities with regard to the electronic structure. The exchange splitting leads to a set of majority and minority spin bands which are non-rigidly shifted with respect to each other. In addition, spin-orbit coupling introduces hybridization points [8, 9] with mixed spin character and avoided crossings. The spin-orbit coupling is also responsible for magnetic dichroic effects [10–12]. However, when probing the electronic structure of cobalt with photoemission, one finds a further, very strong contrast to copper: energy spectra are dominated by rather broad features which possibly hide the multitude of electronic bands that one might expect from independent-particle theoretical calculations. The reason for this is the strong electron-electron interaction.

In this work, we apply a highly efficient scheme of spin-integral as well as spin-resolved photoemission to the study of cobalt thin films grown on Cu(001) substrates. The basis of this scheme is momentum-resolved photoemission [13] using an *aberration-corrected double-hemispherical imaging energy analyzer*. This enables the parallel measurement of the photoelectron ensemble emitted into the half-space above the sample. Measurement times are drastically reduced compared to sequential scans of the photoelectron angle (angle-resolved photoemission, ARPES), and given a certain time frame, a larger amount of data can be collected. This is more than just a convenience. For example, the time frame may be limited by undesired adsorption on the clean sample surface, which gradually alters the information yielded by the photoelectrons. It may also be limited by the allocated beam time at a synchrotron. The second ingredient is the *imaging spin detector* based on low-energy electron diffraction (LEED) at a W(100) surface. This extends the parallelized acquisition of momentum distributions to spin-resolved photoemission enabling an unprecedented efficiency four orders of magnitude higher than single-channel methods. The acquisition of *spin-resolved* momentum distributions in a parallelized manner is a very recent achievement [14]. While the instrumental design and realization took place prior to this work, a working recipe for obtaining spin polarized images was developed as a part of this work. The basic idea is to operate the imaging spin filter in a highly spin-sensitive mode and a weakly spin-sensitive mode. The combined information allows to obtain absolute values for one component of the spin polarization vector. Compared to Mott-type spin detectors, only one of two scattering paths is required, which reduces instrumental asymmetries and keeps the experimental challenge of setting up electron optics for reproducible imaging in a manageable limit. The principles of spin- and momentum-resolved photoemission are introduced in **chapter 3**.

An example for the rapid data acquisition by momentum-resolved photoemission are the measurements shown at the beginning of **chapter 4**, which address the Fermi surface of cobalt. A depiction of the 3-dimensional shape of the spin- and orbital resolved fcc Co Fermi surface obtained from an empirical tight-binding calculation was published in 1995 [15]. No picture nearly as complete exists in the experimental context. Up to now, only a few cross-sections of the Fermi surface of crystalline cobalt have been probed. Three examples of the existing literature illustrate the current state: (1) A sequential measurement of a two-dimensional Fermi surface cross-section of Co on Cu(111) and hcp Co using He-I radiation was reported in [16]. The angular scan was realized by a computer-controlled two-axis sample goniometer. Recording times were in the range of 30 minutes to 24 hours, depending on the photoelectron yield. In this mode, each photoemission intensity value corresponds to a different incidence of light with respect to the crystal axes of the sample, contrary to the momentum-resolved photoemission experiments in this work. (2) Another attempt of probing the Fermi surface of a Co thin film on Cu(001) with photon energies in the range $21\,\mathrm{eV} < h\nu < 45\,\mathrm{eV}$ is reported in [17] – with rather limited success owing to low resolution, low statistics and a doubtful interpretation in ascribing most of the features to final states. In the present work, we did not find any indication of final-state associated resonances in the photon energy range $35\,\mathrm{eV} < h\nu < 200\,\mathrm{eV}$. (3) More recently, two perpendicular cross sections of the Fermi surface of hcp cobalt have been measured at SPRING-8 with soft x-ray radiation ($h\nu = 570\,\mathrm{eV}$) [18]. The $k_{\parallel,y}$ axis is sampled in steps of $0.1\mathrm{\AA}^{-1}$ resulting in a rather coarse-grained, low-detail image of the Fermi surface contour.

In this work, the Fermi surface of an fcc cobalt thin film on Cu(001) was probed by scanning the volume-equivalent of an entire Brillouin zone in 165 steps for the perpendicular direction (section 4.1). Additionally, we present, for the first time, two spin-resolved cross-sections of the Fermi surface (figures 4.9(a), 4.15(a), 7.2) obtained at two different photon energies.

Further experiments presented in chapter 4 probe two areas of the fcc Brillouin zone which contain the highest-symmetrical axes or are close to such axes. The experimental access to the electron spin has the obvious advantage, that majority and minority spin bands are experimentally identified instead of relying on a comparison with theoretically calculated band structures. As mentioned before, a certain degree of spin hybridization is rather the rule than the exception, such that electronic states in the valence band region are usually not $\pm100\%$ spin-polarized. On the other hand, photoelectrons originating from a fully spin-polarized state inevitably mix with a less polarized background. Knowing the spin polarization on continous momentum distributions provides a way of disentangling the background from the spin-polarized peak. Such examples are shown in section 4.2.2. Measuring the influence of relative orientation of the magnetisation with respect to the incidence of light on the spin polarization enables us to estimate the magnitudes of exchange-related spin polarization and spin-orbit related spin polarization (see section 4.2.3).

In section 4.4, two-photon photoemission is employed to access unoccupied parts of the electronic structure. Previously, spin-polarized inverse photoemission enabled a separation of majority and minority quantum well states derived from an unoccupied sp-band [19]. More recently, the same electronic states have been investigated with two-photon photoemission [20], and the observation of linear and circular magnetic dichroism proved the presence of spin-orbit coupling. The spin polarization of photoelectrons emitted in those two-photon processes turned out to be strongly positive [21] and lacked any trace of the minority quantum well state. This was explained by $k_{||}$-conserving transitions that favour majority spin initial states, and secondly, the shorter lifetime of minority spin excited electrons. We extend these investigations to off-normal photoemission, where the chance is much higher to detect a channel for excitation of minority spin electrons via the quantum well state.

A discussion of our experimental results *in view of the electron correlation* is given in **chapter 5**. Electron correlation is an actively studied field of physics and is known to play a role in various interesting phenomena and material classes, such as high-temperature superconductivity, Mott insulators and heavy-fermion systems.

It is furthermore an integral part in the theory of elemental ferromagnetic materials. The valence electronic structure is considered to contain delocalized (sp-band) electrons with wave-like character as well as electrons localized to the atom positions, which behave particle-like (d-band electrons). Both sorts of electrons play a role in the occurence of ferromagnetic order [22]. The localized electrons are more heavily affected by the Coulomb interaction with other electrons than the delocalized electrons. A simple model to describe the mutual repulsion of localized electrons sitting on the same atom is the Hubbard model [23], which is frequently used in the description of strongly-correlated electron systems. The presence of a ferromagnetic ground state (at elevated temperatures) has been shown to require a minimum strength of on-site Coulomb interaction [24]. A theoretical work dedicated to fcc cobalt [25] which made use of the Hubbard model, obtained plausible values of the magnetic moment and of the Curie temperature by choosing the on-site Coulomb interaction energy U=3.5 eV and the on-site interband exchange energy J=0.3 eV.

The effect of electron correlation on the single electron is described by the complex self-energy, where the real part corresponds to the change in energy and the imaginary part to the finite lifetime caused by the electron correlation. Inclusion of the self energy turns a typical band structure plot with sharp lines into a spectral density with "smeared out" lines very similar to real photoemission spectra. Indeed, the imaginary part of the self energy can be evaluated from the linewidth in photoemission spectra. On the other hand, the self energy can be calculated by various theoretical approaches to correlated electron systems, such as dynamical mean-field theory (DMFT [26]) and 3-body scattering theory (3BS [27]). Such theories are in active development and require experimental data to be tested against. The self energy can be written as a function of energy and wave vector. One of the current questions in theory is, how to treat the dependence on the wave vector [28], since the majority of existing calculations consider a "local" self energy which depends only on the binding energy. An interesting question is, whether such an explicit dependence on the wave vector can be proven by experiment. For this purpose, momentum-resolved photoemission is especially helpful, since it covers continous areas of the wave-vector space.

The evaluation of the imaginary part of the self energy from photoemission data and a comparison to current theoretical results is presented in section 5.2.

Chapter 2

Theoretical background

Photoemission is widely used to study the surfaces and the bulk of a variety of materials, such as transition metals [29], ferromagnetic systems [30], superconductors [31, 32], semiconductors [33, 34], band insulators with conducting surfaces, adsorbates on surfaces [35] or thin films on substrates as well as 2-dimensional systems such as graphene [36], or even exotic cases like quasicrystals [37]. The surface sensitivity depends on the photon energy. A great number of experimental works on different materials have shown that the dependence of the inelastic mean free path of electrons on their kinetic energy follows a common trend (a universal curve) [38]. The highest surface sensitivity is found around 40 eV with a probing depth of $\approx 5\text{Å}$. For the x-ray range above 1 keV, the probing depth (>1 nm) is typically sufficient to probe several atomic layers of the sample, while for very low energies (1 eV to 10 eV), it may be as large as 100 nm. With the low pressures achievable today in vacuum chambers, it's possible to prepare a clean surface and maintain it for several hours, so that photoemission with highly surface-sensitive photon energies probes the first few atomic layers of crystalline surfaces and is not too much affected by adsorbates.

2.1 The photoemission process

In order to summarize the basic theoretical formulation of the photoemission process, we start with the quantum-mechanical operator describing the electromagnetic radiation associated with the photon. After that we will go into detail about the initial and final states of photoemission. The photoemission process is driven [39] by the interaction of the solid with a dynamic electromagnetic wave field $\vec{A}(\vec{r}, t)$. The time-dependent (non-relativistic) Schrödinger equation, describing the evolution of the wave function ψ of the electrons in a solid reads:

$$i\hbar\frac{\partial}{\partial t}\psi(\vec{r},t) = \left[\frac{1}{2m}\left(\frac{\hbar}{i}\nabla - \frac{e}{c}\vec{A}(\vec{r},t)\right)^2 + V(\vec{r})\right]\psi(\vec{r},t) = H\,\psi(\vec{r},t) \qquad (2.1)$$

If ψ is a single-electron wave function, the solution describes the dynamics of this electron under the influence of the electromagnetic radiation.

In the dipole approximation, the $\mathcal{A}^2$ term is neglected, and the term in the parentheses reduces to the kinetic energy and a term proportional to $(\nabla\cdot\vec{A}+\vec{A}\cdot\nabla)$, where the gradient of the vector potential $\nabla \cdot \vec{\mathcal{A}}$ contributes only at the surface and the term $\vec{\mathcal{A}} \cdot \nabla$ contributes both at the surface and in the bulk [40]. Focussing on the photoemission processes in the bulk, the effect of the electromagnetic field is represented by the operator $-\frac{eh}{imc}\vec{\mathcal{A}} \cdot \nabla$. The transition rate $w_{\mathrm{f,i}}$ from an initial state i to a final state f using first order time-dependent perturbation theory then reads:

$$w_{\mathrm{f,i}} \propto \frac{2\pi e}{imc} \left| \langle f \, | \, \vec{\mathcal{A}} \cdot \nabla \, | \, i \rangle \right|^2 \delta(E_f - E_i - h\nu) \tag{2.2}$$

In real solids, the initial and final states of a photoemission process are more complex than a single electron occupying a lower or higher energy level. In general, the electrons interact with each other during the photoemission process. Therefore, one has to replace the one-particle electronic initial (final) states by many-particle wave functions, that describe the whole ensemble of electrons in its ground state (excited state). The energy $h\nu$ of the photon is absorbed by the electron ensemble and may lead to the emission of a photoelectron, while the remaining electron system reacts to the missing electron charge (the photohole) [41]. If we compare this to the independent-electron picture, the initial state electron occupying a single-particle energy level would absorb the photon energy, be emitted as a photoelectron and detected – the refilling of the original level would take place in a separate process and not influence the energetics of the photoelectron. Contrarily, in the interacting electron system, the emission process and the relaxation process cannot be separated. The energy of the absorbed photon is distributed onto the emitted electron and the remaining electron ensemble. In a solid, the proper theoretical description involves a spectral function $A(\vec{k}, E)$, such that the photocurrent I reads [41]

$$I \propto \sum_{f,i,\vec{k}} \left| \left\langle \phi_{f,E_{\mathrm{kin}}} \left| r \right| \phi_{i,\vec{k}} \right\rangle \right|^2 A(\vec{k}, E) \tag{2.3}$$

The spectral function describes the probability of removing (adding) an electron with energy E and wave vector $\vec{k}$ from (to) the interacting N-electron system. In case of a noninteracting system, it has a δ-like infinite density at the energies and wave vectors of the single-electron bands. For interacting electrons, a complex-valued self energy $\Sigma(\vec{k}, E)$ is introduced:

$$A(\vec{k}, E) = \frac{1}{\pi} \frac{\Im(\Sigma(\vec{k}, E))}{\left(E - E_{\vec{k}}^0 - \Re(\Sigma(\vec{k}, E)) \right)^2 + \left(\Im(\Sigma(\vec{k}, E)) \right)^2} \tag{2.4}$$

Here, $E_{\vec{k}}^0$ are the hypothetical energy bands of the non-interacting electrons. The energy of the quasiparticle is shifted relative to $E_{\vec{k}}^0$ by the real part of the self energy. The lifetime of the photohole - limited by many-body interactions - results in energy broadening. This broadening is quantified by the imaginary part of the self energy.

In the one-step model of photoemission developed by Pendry [42], the final state $|f\rangle$ is a time-reversed LEED state. On the vacuum side, it consists of a plane wave with the $\vec{k}$-vector pointing in a particular direction of the half-sphere above the sample. At the surface, it is matched to the electronic states of the sample by requiring continuous amplitude and continuous first derivative of the wave function. This matching condition results in a conservation of the surface-parallel wave vector components $k_{||} = (k_{||,x}, k_{||,y})$, since the translational symmetry along the surface-parallel directions is not reduced when the photoelectron propagates from the sample to the vacuum. The $k_{||}$ conservation holds as long as the half-infinite ideal crystal is a good approximation of the sample. If the vacuum plane wave can be matched to bulk electronic states (Bloch states) at the final state energy, the photoelectron wave function extends into the sample and the surface sensitivity is determined by the escape depth of the photoelectron due to inelastic scattering. If there are no matching bulk electronic states at the final state energy, the wave function decays exponentially into the sample. This kind of wave function is denoted as evanescent state [43] and also contributes to the photoemission current. The corresponding part of the photoemission signal may have a different surface sensitivity determined by the penetration depth of the wave function.

In the surface normal direction, the translational symmetry is violated by the sample surface and hence the $k_\perp$ component is not conserved. In principle, Bloch waves with any value of the $k_\perp$ component can couple to the vacuum part of the final state, but the amplitude will be strongest for particular values of $k_\perp$. Typically, electronic states in a solid, that are several tens of eVs above the vacuum level, have a similar dispersion as the free electron. The difference is that band gaps may occur at the Brillouin zone boundary and that the dispersion relation is (rigidly) shifted to lower energies. The latter is due to the difference of the electric potential in a solid with respect to the vacuum. This difference - the so-called inner potential - is equivalent to the height of the surface barrier (plus a contribution from exchange-interaction between electrons) [44].

When recording a momentum- and energy-resolved photoemission spectrum, one will find intensity maxima at specific values of $k_{||}$ and energy E_{kin}. These can be produced by direct transitions between valence electronic bands and unoccupied bands at the final state energy. In terms of the spectral density this means, that either the final state density or the initial state density should have a local maxima. Since we can only explicitly select $k_{||}$, E_{kin} and $h\nu$, but not $k_\perp$ in the photoemission measurement, transitions involving $A(k_{||}, k_\perp, E_i)$ and $A(k_{||}, k_\perp, E_i + h\nu)$ are possible for all $k_\perp$ and the one, where the product of both spectral density values is maximum will contribute most to the photoemission signal (if the transition matrix element between the electronic states does not vanish). Hence, as far as transition rates are determined by the bulk spectral density, photoemission intensity maxima can be associated with wave vectors $\vec{k} = (k_{||}, k_\perp)$, where $k_\perp(E_f, k_{||})$ is determined by direct transitions to the unoccupied electronic band structure and $k_{||}$ selected by the experiment.

2.1.1   The free-electron final state model

For photoemission resonances that involve transitions from bulk electronic states, a very common approximation to infer the value of the perpendicular wave vector component $k_\perp$ is to replace the actual final state band dispersion by the dispersion of the free electron.

$$E_f(\vec{k}) = E_f(k) = \frac{\hbar^2 k^2}{2m} + (E_{\text{vac}} - U_{\text{i}}) \qquad (2.5)$$

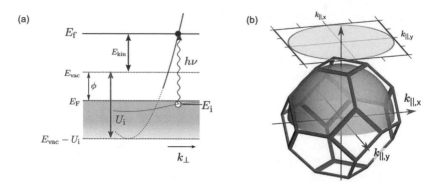

Figure 2.1: (a) The dispersion of the unoccupied bulk electronic bands is approximated by a free-electron dispersion (blue curve) with minimum at $E_{vac} - U_i$. The filled (empty) circle indicates a photoelectron (photohole) created by a direct transition from a valence electronic state to this unoccupied bulk electronic state. (b) $\vec{k}$-sphere generated by the condition $E_f(\vec{k}_{||}, k_\perp) = \text{const}$, where the constant increases with the kinetic energy of the photoelectron. Assuming a free-electron dispersion of the final state bands, photoemission current produced by interband direct transitions probes bulk electronic states with Bloch wave vectors $\vec{k}$ lying on the orange spherical surface. The photoemission momentum distributions presented in this work show projections into the $k_{||}$ plane as sketched by the coordinate frame on top of the Brillouin zone.

This describes a sphere in k-space and a parabola for the energy as a function of the absolute value of the wave vector k. The photoelectron may be scattered by the crystal, which alters the wave vector $\vec{k}$ by a reciprocal lattice vector $\vec{G}$:

$$E_f(\vec{k} + \vec{G}) = \frac{\hbar^2}{2m}\left((k_{||,x} + G_{||,x})^2 + (k_{||,y} + G_{||,y})^2 + (k_\perp + G_\perp)^2\right) + (E_{vac} - U_i) \quad (2.6)$$

In this case, the k-sphere is centered around $-\vec{G}$. The energetic position of the parabola minimum is $E_f(0) = E_{vac} - U_i$ where U_i is called the inner potential, see figure 2.1(a). In this work, we present momentum distribution maps for various values of E_f. In the reciprocal lattice space of wave vectors $\vec{k}$, the condition E_f=const constitutes a $\vec{k}$-sphere, intersecting the Brillouin zone. The radius of the spherical constant-energy surface is given by $\frac{\sqrt{2m}}{\hbar}\sqrt{E_{kin} + U_i}$, which for easy numerical evaluation can be written:

$$r_k[\text{Å}^{-1}] \approx 0.5123\sqrt{(E_{kin} + U_i)[\text{eV}]} \quad (2.7)$$

One alternative for obtaining $k_\perp(E_{kin}, k_{||})$ has been discussed by Strocov [45], who stated that VLEED probes the same states that are the final states in photoemission. Therefore, final state bands can be fit to the critical points detected in VLEED. However, it was noted, that some of the most important final state bands in photoemission are not visible in VLEED. Deviations from the free-electron dispersion occur primarily at the Brillouin zone border [46]. Secondly, calculations of the final state band structures can be performed.

2.2 Density functional theory and Local density approximation

When describing a many-body (fermionic) system by a many-body wave function, it may be formulated in terms of the position of every particle, which becomes numerically unmanageable even for small clusters of a crystalline solid. In density functional theory this complexity is reduced by exploiting the Hohenberg-Kohn theorem which states that the ground state wave function is a functional of the ground-state electron density ρ [47]. This means, that ρ defines the electron system completely, all ground-state observables are functionals of ρ, too. Instead of solving the Schroedinger equation, the total energy is written as a functional of the density and its minimum (corresponding to the ground state) is found by a variation of the density $\rho(\vec{r})$. One can split up this energy functional into several terms [48]:

$$E[\rho] = T_{\mathrm{s}}[\rho] + \int \left(V_{\mathrm{ext}}(\vec{r})\rho(\vec{r}) + \frac{1}{2} V_{\mathrm{H}}[\rho(\vec{r})]\rho(\vec{r}) \right) d^3r + E_{\mathrm{xc}}[\rho] \tag{2.8}$$

where T_{s} is the kinetic energy corresponding to an independent-particle system with density ρ, V_{ext} is the potential energy from electron-ion Coulomb interaction, V_{H} is the Hartree potential given by $e^2 \int \frac{\rho(\vec{r}')}{|\vec{r} - \vec{r}'|} d^3r'$ and E_{xc} is the exchange-correlation energy. The exact form of E_{xc} is unknown and one of the main questions of DFT is how to find a succesful approximation.

The local density approximation (LDA) chooses a special form of E_{xc}: Both exchange and correlation energy are taken from a homogeneous electron liquid, where the former is given (as a functional) by [47, 49]

$$E_n^{\mathrm{LDA}}[\rho] = -\frac{3q^2}{4} \left(\frac{3}{\pi} \right)^{1/3} \int d^3r \rho(\vec{r})^{4/3} \tag{2.9}$$

and the correlation part by numerical data from Quantum Monte Carlo calculations made by Ceperley and Alder [47, 50]. This means, the energy functionals correspond to the homogeneous electron liquid, but they are evaluated for the non-homogeneous density of the actual system being treated.

2.3 Strongly correlated electron systems

Electron-electron correlation in the solid state can lead to a variety of interesting physical phenomena and material properties such as Mott-type metal-insulator transitions [51, 52], high-temperature superconductivity [53], or heavy fermion behaviour [54]. These systems have in common that a theoretical description within a single-electron picture does not predict the true properties. In a single-particle description, the electron-electron interaction may be considered to some extent in the form of a time-independent effective potential. However, strong Coulomb repulsion between two spatially adjacent electrons may lead to insulating behaviour (Mott insulator) in cases where a single-electron theory would find metallic conductivity with a non-vanishing density of states at the Fermi level.

An attempt to describe the electron correlations is the Hubbard model [23]. The author focussed on the description of narrow energy bands (where Wannier functions are stronger localized at the atom positions than for highly dispersive bands spanning a large energy range). The terms, that were included in the Hamiltonian, describe electron hopping between neighbouring atoms and Coulomb repulsion of two electrons with opposite spin on the same atomic site. By increasing the second term, the model reproduces the formation of a band gap at the Fermi level. In fact, the strength of correlation is found for a wide variety of materials to increase with the ratio of the on-site Coulomb repulsion to the band width (e.g. figure 1 in [55]).

While one might expect strong correlation effects in metals with a high density of conducting electrons, this is often not the case. One reason is that the rate of inelastic electron-electron scattering in metals is largely reduced due to screening of the Coulomb field by the high-density electron gas [56]. Another reason is that the kinetic energy is larger than the correlation energy which consists of the Coulomb interaction between electrons[1]. Landau [59–61] found that the collective behaviour after exciting the electron system by a photon can be separated into (1) high-energetic collective oscillations of the electron sea (plasmons) and (2) low-energetic excitation of a single electron accompanied by some response of the remaining electrons. The latter was found to yield an excited electron with relatively well-defined energy and two characteristics: (1) The excitation energy is shifted ("renormalized") with respect to an independent-electron description. (2) the energy spectrum of the excited electron has an increased linewidth. These modifications are quantified by real and imaginary part of the self-energy and the excited electron is denoted as a quasiparticle.

2.4   Probing the spectral function $A(\vec{k},E)$ by photoemission

As mentioned in section 2.1, the spectral function $A(\vec{k}, E)$ probed by photoemission contains information about the independent-particle energies $E_{\vec{k}}^0$ as well as their renormalization ($\Re(\Sigma)$) and lifetime broadening ($\Im(\Sigma)$) due to many-body interactions. During the photoemission process, the finite lifetime of both photohole and photoelectron influence the linewidth that is measured in energy or momentum spectra.

The photocurrent as a function of kinetic energy and parallel momentum is proportional to a convolution of the spectral density $A(\vec{k}, E)$ evaluated for the initial state i and the final state f [62]:

$$I(E, \vec{k}_\parallel) \propto \int dk_\perp\, A(\vec{k}_\parallel, k_\perp, E - h\nu)\, A(\vec{k}_\parallel, k_\perp, E) \tag{2.10}$$

In the following, a simplified example demonstrates how the dispersion of initial and final state bands alter the linewidths, that can be observed in the principal measurement

---

[1]The kinetic energy may be estimated by $E_{\mathrm{kin}} = 2.21/r_s^2$ [Ryd] and the Coulomb energy by the Hartree-Fock potential energy $0.916/r_s$ [Ryd] or by the correlation energy $\approx (0.0622 r_s - 0.096)$[Ryd] [57] with typical values of the Wigner-Seitz radius r_s for metals ranging from 2 to 6 bohr radii (copper: $r_s = 2.62$ [58]). Hence, for lower densities the Coulomb term increases more strongly than the kinetic energy.

modes, with respect to the imaginary part of the self-energy. The following spectral function $A(\vec{k}, E)$ describes a broadened initial state band and a non-broadened final state.

$$A(\vec{k}_{||}, k_\perp, E) = \begin{cases} A^0_{i, \Sigma_i}(\vec{k}_{||}, k_\perp, E) \\[2mm] A^0_f(\vec{k}_{||}, k_\perp, E) \end{cases}$$

$$= \begin{cases} \dfrac{\Im(\Sigma_i)}{(E - E^0_i(\vec{k}_{||}, k_\perp) - \Re(\Sigma_i))^2 + \Im(\Sigma_i)^2} & (E < E_F) \\[4mm] \delta(E - E^0_f(\vec{k}_{||}, k_\perp)) & (E > E_F) \end{cases} \qquad (2.11)$$

For $E > E_F$ and $E - h\nu < E_F$, equation (2.10) evaluates to

$$I(E, \vec{k}_{||}) \propto \int dk_\perp \, A^0_{i, \Sigma_i}(\vec{k}_{||}, k_\perp, E - h\nu) \, \delta(E - E^0_f(\vec{k}_{||}, k_\perp)) \qquad (2.12)$$

Defining $k^0_{\perp, f}(E, k_{||})$ as the inverted function of $E^0_f(\vec{k}_{||}, k_\perp)$, the integration yields

$$I(E, \vec{k}_{||}) \propto A^0_{i, \Sigma_i}(\vec{k}_{||}, k^0_{\perp, f}(E, k_{||}), E - h\nu) \qquad (2.13)$$

This means, the final state band enforces a certain path $k^0_{\perp, f}(E, k_{||})$ through the spectral density in the initial state energy region, when we vary either the final state energy E or the parallel momentum component. Probing the spectral density along a particular path results in the width of peaks that we measure in momentum and energy-resolved spectra.

In this work we have frequently recorded momentum-resolved images for a continous range of kinetic energies keeping the photon energy constant. In the resulting dataset, the widths of photoemission intensity peaks can be evaluated either along the energy axis or along a particular direction in the $(k_{||,x}, k_{||,y})$-map. In the following we derive the linewidths in energy spectra at constant $(k_{||,x}, k_{||,y})$ and momentum-resolved spectra at constant final state energy E_f. Figure 2.2(a) shows how the path of final state bands (shifted by $-h\nu$) through the initial state spectral density leads to the measured linewidth in an energy spectrum. This is shown for two photon energies $h\nu_1, h\nu_2$ which in this case probe the initial state spectral density at the same wave vector and for the same initial state energy range, but with two different slopes of the final state band. The measured linewidth for $h\nu_1$ yields almost the same value as $2\Im(\Sigma_i)$ – the full width at half maximum of the quasiparticle band – while the measured linewidth for $h\nu_2$ is significantly increased due to the similar slopes of $dE^0_i/dk_\perp$ and $dE^0_f/dk_\perp$. To calculate the measured linewidth in an energy spectrum, we start from the energy of the photoemission peak at $k^p_\perp$, i.e. $E^0_i(k^p_\perp) = E^0_f(k^p_\perp) - h\nu$. The half value of the peak density is probed by the final state band at deviating $k_\perp$, when $E^0_f(k^p_\perp \pm \Delta_\pm k_\perp) - E^0_i(k^p_\perp \pm \Delta_\pm k_\perp) = \pm\Im(\Sigma_i)$. That means, the final state band energy has to deviate *relative to the initial state band energy* by an amount of $\pm\Im(\Sigma_i)$. In first order approximation ($\Delta_\pm k_\perp$ then become equal):

$$\frac{d(E^0_f - E^0_i)}{dk_\perp} 2\Delta k_\perp = 2\Im(\Sigma_i) \qquad (2.14)$$

The corresponding energy interval W_E that the energy analyzer has to be scanned through is spanned by the dispersion of the final state energy across the same interval $2\Delta k_\perp$:

$$\frac{dE_f^0}{k_\perp} 2\Delta k_\perp = W_E \tag{2.15}$$

Eliminating $2\Delta k_\perp$ in equations (2.14) and (2.15) yields

$$W_E = 2\Im(\Sigma_i)\frac{v_f}{v_f - v_i} = \frac{2\Im(\Sigma_i)}{1 - v_i/v_f} \tag{2.16}$$

where we have introduced $v_i = dE_i^0/dk_\perp$ and $v_f = dE_f^0/dk_\perp$ The latter expression allows to calculate the limit of the final state band (shifted by $-h\nu$) vertically cutting through the initial state density which results in an energy width of $2\Im(\Sigma_i)$ as expected:

$$v_f \to \infty \quad \Rightarrow \quad W_E = 2\Im(\Sigma_i) \tag{2.17}$$

Equation (2.16) is invalid for $v_i = v_f$, since, in case of vanishing first derivative in equation (2.14), the second derivatives have to be included, at least. Smith et al.[63] have calculated the experimental linewidths of photoemission spectra including a finite lifetime of the final state. However, they aim at angle-resolved photoemission spectroscopy (ARPES) , which implies fixed emission angle and therefore, varying $k_{||}$, leading to rather complicated expressions. Equation (2.16) is equivalent to their equation for a so-called energy distribution curve (EDC, $h\nu$=const, E_f is swept) in *normal emission* (the only condition where $k_{||}$ does not vary implicitly in an ARPES measurement)

$$W_E = \frac{2\Im(\Sigma_i)\, v_f - 2\Im(\Sigma_f)\, v_i}{v_f - v_i} \tag{2.18}$$

if we set the lifetime broadening $\Im(\Sigma_f)$ of the final state to zero. Since our photoemission spectrometer provides spectra with constant $k_{||}$ in off-normal emission, as well, they are treated correctly by equation (2.18).

Now, we look at momentum-resolved spectra for $E_f, h\nu = $ const; $E_i = E_f - h\nu = $ const. We assume that line profiles are selected along the $k_{||,x}$ axis, such that $k_{||,y} = $ const. Hence, we plot the constant energy surfaces initial and final state bands in figure 2.2(b) in the $(k_{||,x}, k_\perp)$ plane. For the initial state, the constant-energy surface is implicitly broadened in k-space as a result of the the explicit energy broadening $\Im(\Sigma)$. The final state constant-energy surface probes the resultant spectral density and its peak positions in a particular direction of the $(k_{||,x}, k_\perp)$ plane, which can be expressed by the tangent vector $\vec{e}_f$:

$$\vec{e}_f = \begin{pmatrix} 1 \\ 0 \\ \frac{dk_{\perp,f}^0}{dk_{||}} \end{pmatrix} \quad |\vec{e}_f| = \sqrt{1 + \left(\frac{dk_{\perp,f}^0}{dk_{||}}\right)^2}^{\,-1} \tag{2.19}$$

Here, $k_{\perp,f}^0 = k_{\perp,f}^0(k_{||,x}, k_{||,y})$ is a parametrization of the final state constant-energy surface.

Along the straight line that points in the direction of $\vec{e}_f$ and goes through an intersection point $\vec{k}^p$, there are two points $\vec{k}_\pm^p$ on either side of $\vec{k}^p$, where the initial state band energy E_i^0 deviates from the selected $E_i = E_f - h\nu$ by an amount of $\Im(\Sigma)$. The distance

$\kappa = |\vec{k}_+^p - \vec{k}_-^p|$ equals the FWHM of a spectral density line profile along the $\vec{e}_f$ axis. Since the rate of change in initial state band energy E_i^0 is given in first order approximation by the k-gradient of the initial state band energy E_i^0 projected onto $\vec{e}_f/|\vec{e}_f|$, we can write:

$$\vec{\nabla}_k E_i^0 \cdot \frac{\vec{e}_f}{|\vec{e}_f|} \approx \frac{2\Im(\Sigma)}{\kappa} \quad \Rightarrow \quad \kappa \approx \frac{2\Im(\Sigma)}{\vec{\nabla}_k E_i^0 \cdot \vec{e}_f/|\vec{e}_f|} \tag{2.20}$$

The quantities $\vec{e}_f$, κ, $\vec{\nabla}_k E_i^0$ are illustrated in figure 2.2(c). The projection onto the $k_{||,x}$ axis is obtained by multiplying κ with the $k_{||,x}$ component of $\frac{\vec{e}_f}{|\vec{e}_f|}$.

$$W_{k_{||,x}} = \frac{\kappa}{|\vec{e}_f|} \tag{2.21}$$

Inserting equations (2.19) and (2.20) into (2.21), we obtain

$$\boxed{W_{k_{||,x}} = \frac{2\Im(\Sigma)}{\frac{dE_i^0}{dk_{||,x}} + \frac{dE_i^0}{dk_\perp}\frac{dk_{\perp,f}^{0,}}{dk_{||,x}}}} \tag{2.22}$$

Analogous relations taking additionally into account the final-state lifetime broadening, have been derived and discussed in a number of publications [62–64]. Lifetimes of low-energy electrons in metals (corresponding to the final-states in ultra-violet photoemission spectroscopy) have been theoretically treated by Echenique [65]. Spin-dependent lifetimes of excited electrons in transition metals have been investigated by theory and experiment in [66]. Citing two numerical example from literature, Sanchez-Royo et al [64] have determined a lifetime broadening (HWHM) of $3.5 \pm 0.6\,\mathrm{eV}$ for electrons excited to $E_F + 32\,\mathrm{eV}$ in Ag(111). Pendry and coworkers used an inverse electron lifetime of $4\,\mathrm{eV}$ for copper and energies in the range $E_F + 10.8 \ldots E_F + 16.8$. For the jellium model (electrons in a homogeneous positively charged background), the following equation describes energy dependence of the lifetime of excited("hot") electrons [56, 65][2]

$$\tau = 263\, r_s^{-5/2}\, (E - E_F)^{-2}\; \mathrm{fs\,eV^2} \tag{2.23}$$

In real solids, the self energy Σ of photohole and photoelectron quasiparticles is not constant. The imaginary part (inverse lifetime) vanishes at the Fermi level (if only electron-electron scattering is taken into account) and increases towards lower and higher energies. From the Fermi liquid model of electrons in metals, one obtains $\Im(\Sigma) \propto (E - E_F)^2$ [67, 68] with rather small prefactors in case of weakly interacting electron systems (e.g. copper). However, for the *ferromagnetic* transition metals, the broadening due to $\Im(\Sigma)$ is considerably stronger and does not follow a simple parabolic form. Pendry et al. [69] discussed the stronger broadening of photoemission from (paramagnetic) nickel compared to copper in view of the photohole lifetimes that are shorter in nickel. The photohole decays via Auger processes, which proceed at a higher rate, the higher the spectral density is in the vicinity of the Fermi level. For nickel (as well as for cobalt) the spectral density above and below the Fermi level is strongly enhanced by the d-bands, while for copper

---

[2]inverse lifetime in eV: $\hbar/(2\tau) = 658\mathrm{meV\,fs}/(2\tau)$

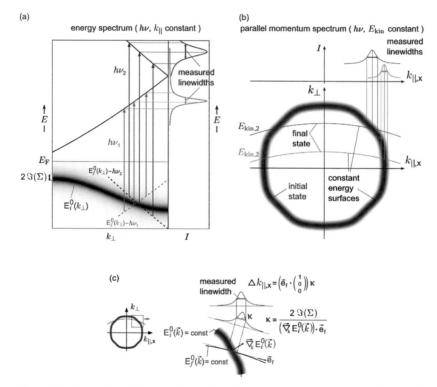

Figure 2.2: Illustration of the direct transitions involved (a) in energy spectra at constant $k_{||,x}, k_{||,y}, h\nu$; (b) in momentum-resolved spectra at constant $E_f, E_i, h\nu, k_{||,y}$. The energy broadening $2\Im(\Sigma)$ of the initial state spectral density causes peaks with finite widths for case (a) as well as case (b). See text for more details.

the d-bands are $\approx 2\ eV$ below the Fermi level. More recent data regarding the lifetime-broadening based on theoretical approaches can be found in ([70–73]). Experimental works have been carried out for iron [74] (see reference 12 therein, data assembled from multiple experimental works) and Ni(110) [75]. In principle, the self-energy Σ may depend on energy E and wave vector $\vec{k}$ of the electronic quasiparticle, symmetry character of the wave function (e.g. calculated separately in [76]) as well as electron spin (where large differences occur in ferromagnetic systems). The correct description of many-body interactions leading to the electron self-energy is an active research field with some of the recent innovations being dynamical mean-field theory (DMFT) [26, 77–79] and nonlocal (wave vector-dependent) treatment of the self-energy [28, 80–82]. In this work, we shall attempt to probe $\Im(\Sigma(\vec{k}, E))$ at several $\vec{k}$ and E for Co/Cu(001).

2.5 Spin-orbit coupling

In 1927, Pauli described an additional angular momentum degree of freedom of the electron [83] which could assume either of two discrete values $\pm \hbar/2$. This became to be known as the electron spin and was recognised as a relativistic effect of quantum mechanics by Dirac [84, 85] in 1928. The related phenomenon of spin-orbit coupling is another consequence from the relativistic treatment as shown below (following Kessler [86]):

We start from the relativistic energy law

$$E^2 = c^2 p^2 + m^2 c^4 \tag{2.24}$$

where E is the energy, c the speed of light, $\vec{p}$ the momentum and m the rest mass and substitute the Hamilton operator $E \rightarrow H = i\hbar \frac{\partial}{\partial t}$ and the momentum operator $\vec{p} \rightarrow -i\hbar \nabla$ to change from a classical description to a quantum mechanical one. This leads to the occurence of a second derivative with respect to time, posing several conceptual problems. Dirac found a way to linearize this equation at the cost of introducing matrix coefficients and arrived at

$$\left[i\hbar \frac{\partial}{\partial t} + i\hbar c \left(\alpha_x \frac{\partial}{\partial x} + \alpha_y \frac{\partial}{\partial y} + \alpha_z \frac{\partial}{\partial z} \right) - \beta mc^2 \right] \psi = 0 \tag{2.25}$$

where $\alpha_x, \alpha_y, \alpha_z, \beta$ are 4×4-matrices and ψ is a wave function of the form $(\psi_1, \psi_2, \psi_3, \psi_4)^{\mathrm{T}}$. Two remarkable consequences follow:

- The angular momentum $\vec{l}$ does not commute with the Hamiltonian for radial symmetric potentials. However, an additional angular momentum $\vec{s}$ can be constructed, so that the sum of both $\vec{l} + \vec{s}$ commutes with H and is therefore a conserved quantity. $\vec{s}$ is found to be the observable of the electron spin.

- Introducing electromagnetic fields[3] by substituting $\vec{p} \rightarrow \vec{p} - \frac{\varepsilon}{c}\mathcal{A}$ and $H \rightarrow H - \varepsilon\phi$ and

---

[3] symbols: vector potential $\mathcal{A}$ as defined by $\vec{B} = \nabla \times \vec{A}$, electric potential ϕ, electric permittivity ε, magnetic field $\vec{B}$, electric field $\vec{\mathcal{E}}$, electron energy W excluding the rest energy $m_0 c^2$

approximating $E_{\text{kin}}, V(\vec{r}) \ll mc^2$, the nonlinear equation (2.24) leads to

$$\left[\frac{1}{2m}\left(\vec{p} - \frac{\varepsilon}{c}\vec{A}\right)^2 + \varepsilon\phi - \frac{\varepsilon\hbar}{2mc}\vec{\sigma}\,\vec{B} + i\frac{\varepsilon\hbar}{4m^2c^2}\vec{\mathcal{E}}\,\vec{p} - \frac{\varepsilon\hbar}{4m^2c^2}\vec{\sigma}(\vec{\mathcal{E}}\times\vec{p})\right]\psi = W\psi$$

(2.26)

The first two terms do not differ from the Schroedinger equation for a particle in an electromagnetic field. The third term, where $\vec{\sigma}$ is related to the spin observable $\vec{s}$ by $\vec{s} = (\hbar/2)\,\vec{\sigma}$, represents the interaction energy of the electron spin with an external magnetic field. The fifth term describes spin-orbit coupling.

In the non-relativistic limit, two components of ψ can be neglected and a two-component spinor can be used to describe the electron spin [87]. Its measurement along a defined spin quantization axis is expressed by one of the three Pauli matrix operators

$$\sigma_x = \begin{pmatrix} 0 & 1 \\ 1 & 0 \end{pmatrix} \quad \sigma_y = \begin{pmatrix} 0 & -i \\ i & 0 \end{pmatrix} \quad \sigma_z = \begin{pmatrix} 1 & 0 \\ 0 & -1 \end{pmatrix} \quad \vec{\sigma} = \begin{pmatrix} \sigma_x \\ \sigma_y \\ \sigma_z \end{pmatrix}$$

(2.27)

The spin-orbit term contains a scalar product between the Pauli matrix vector $\vec{\sigma}$ (the observable of the spin polarization vector $\vec{P} = (P_x, P_y, P_z)^{\mathsf{T}}$) and the term $\vec{\mathcal{E}}\times\vec{p} = \nabla V\times\vec{p}$ which depends on the orbital motion of the electron. In case of a radial symmetric potential (as for an electron in an atom), $\nabla V(\vec{r}) = \frac{\partial V}{\partial r}\vec{r}/r$. In that case, the spin-orbit term is proportional to $\vec{\sigma}\vec{L}$, which means, it depends on the relative orientation of spin and angular momentum.

In the absence of spin-orbit coupling, the wave functions of an electron in a solid can be written as a composition of a spatial and an independent spin part [88]. The symmetries of the spatial part are described by irreducible representations of a point group of the Bloch vector $\vec{k}$, also termed "single group". In contrast, when spin-orbit coupling is present, the electron wave functions have the form

$$\psi(\vec{k}) = |a\rangle\,|\uparrow\rangle + |b\rangle\,|\downarrow\rangle,$$

(2.28)

where $|a\rangle$, $|b\rangle$ are different orbital wave functions, $|\uparrow\rangle$, $|\downarrow\rangle$ are orthogonal spin states of some quantization axis. Then, the symmetries of the electron wavefunctions have to be described by irreducible representations of double groups which have the double number of symmetry elements compared to the corresponding single group [89] (all elements of the single group plus all elements combined with a rotation about 2π, since that rotation changes the sign of a spinor). The double group contains additional irreducible representations, which mix the spatial symmetries from the single group. Examples of such hybridization of electronic states have been presented, e.g. for Fe(110) in [90]. As a result, transitions from such electronic states by electric dipole excitation (aligned along crystallographic directions) obey selection rules, that predict a spin polarization effect in some cases (see [88, 91]).

Spin-orbit coupling preserves time reversal symmetry, since both σ and $\vec{p}$ in the fifth term of equation (2.26) change sign when reversing time and therefore the net sign of the product does not change. If no other terms (e.g. magnetic fields) break the time reversal symmetry, the energy eigenvalues of an electron obey $E(\vec{k},\uparrow)=E(-\vec{k},\downarrow)$ (with Bloch wave vectors $\vec{k}$ and spinors $|\uparrow\rangle, |\downarrow\rangle$.) When both time reversal and spatial inversion are symmetry operations of the system, $E(\vec{k},\uparrow)=E(\vec{k},\downarrow)$ follows, i.e. the electronic bands for spin up and spin down have the same energy eigenvalues. Inversion symmetry can be violated either in the bulk of non-centrosymmetric crystals (zinc blende semiconductors with spin-orbit split valence bands [92]) or at crystal surfaces (inversion means $\vec{x} \to -\vec{x}$ and at a surface, $+z$ belongs to the vacuum half space while $-z$ belongs to the crystal half space). An example for the latter is splitting of surface states on Bismuth [93].

2.5.1   Magnetic dichroism

Photoemission measurements on magnetic surfaces can be characterized by the following complete set of experimental parameters:

- angles of incidence θ, φ of the photon with respect to the surface normal (polar angle θ) and to the surface crystallographic axes (azimuth φ). φ defines the optical plane (if $\theta \neq 0$) that is spanned by the surface normal and the wave vector of the light.

- Polarization of the light. In case of fully polarized light, it can be expressed by a two-component complex Jones vector $(E_{0,x}\, e^{i\phi_x},\ E_{0,y}\, e^{i\phi_y})^{\mathrm{T}}$ that contains amplitudes $E_{0,x}, E_{0,y}$ and phase shifts ϕ_x, ϕ_y of the electric field vectors along two orthogonal axes perpendicular to the propagation of the light (z). Special cases are linear polarization with the electric field vector $\vec{E}$ perpendicular to the optical plane (s-polarized) or within the optical plane (p-polarized) and circular polarization where $\vec{E}$ rotates about the direction of propagation.

- Magnetization $\vec{M}$

- momentum $\vec{p}$ of the detected photoelectron

- spin $\hbar/2\vec{\sigma}$ of the photoelectron

Phenomenologically, magnetic dichroism means that the spin-integral photoemission intensity changes after reversal of the magnetization direction ($\vec{M} \to -\vec{M}$).

  This effect occurs when using circular polarized light [88] as well as linear polarized light [91, 94]. An analytical treatment covering (001), (110) and (111) surfaces of cubic lattices is given in [91]. Occurence of magnetic dichroism requires both the presence of spin-orbit coupling and exchange interaction. If in the non-magnetic limit, spin-orbit coupling produces an electron spin polarization component, and this component is parallel to the magnetization, then magnetic linear dichroism will be observed [91]. Magnetic dichroism has been used to identify spin-orbit hybridization points in the electronic structure by looking for plus/minus features in the spectra of the dichroic asymmetry [94].

2.6   Two-photon photoemission

Two-photon photoemission (2PPE) is a nonlinear process, that occurs when a high photon flux (corresponding to high electric field strengths) arrives at the sample surface. The photocurrent is then proportional to the square of the intensity (fourth power of the electric field). The necessary photon flux can be generated in pulsed lasers.

Variants with monochromatic [95–97] or bichromatic [95, 98, 99] radiation have been performed, probing intermediate states in different energy ranges. Two main purposes of this technique are the study of unoccupied electronic states [95, 100, 101] and the study of dynamics by time-dependent photoemission [102–107], for example relaxation processes of excited electrons. In principle, it can also be used to probe primarily initial states [108]. Time-resolved measurements can be realized by sending the laser pulse through a beam splitter and a delay stage where the path lengths, along which each of the split pulses travel, is used to tune the time difference [109]. The first pulse then effects an initial excitation of the electronic system while the delayed second pulse probes the excited state, leading to the emission of a photoelectron.

In this work, the focus was less on dynamics than on studying the unoccupied electronic structure of cobalt thin films *with spin resolution*. Former two-photon photoemission (for $\vec{k}_\parallel = 0$) studies on this material system have shown that the unoccupied electronic structure is accessible with photon energies $h\nu_1 = h\nu_2 = 3.1$ eV for the pump and probe pulses, respectively [8, 21].

The 2PPE process can proceed (simultaneously) in two different ways: (1) a direct two-photon ionization which transfers the electron from the initial to the final state. No intermediate state is required and energy conservation implies that the *sum* energy (i.e. the sum frequency generated in the non-linear process) equals the difference of initial and final state energies $E_f - E_i = h\nu_1 + h\nu_2$; (2) a step-by-step process which populates an intermediate state with one photon and subsequently excites the electron to the final state with the second photon. Here, energy conservation is fulfilled separately for each step [110]. The pump pulse is tuned in *resonance* to the energy difference between allowed initial and intermediate states $E_j(\vec{k}_0) = E_i(\vec{k}_0) + h\nu_1$. This process is sensitive to the intermediate state and, for instance, can be used to trace its dispersion in momentum-resolved and energy-resolved spectra. In [21], it was shown, that such a resonant process may produce photoelectrons with enhanced spin polarization compared to the non-resonant two-photon processes due to spin-selective transitions in the pump excitation step.

Several works have calculated the line shapes of the 2PPE spectra for discrete three-level systems and in the limit of a continous photon field (i.e. laser pulses long compared to the lifetime of the intermediate state), as well as for specific shapes and durations of laser pulses [111, 112]. We quote a simplified expression for the 2PPE lineshape from [110], where a Lorentzian density of states was used for the initial state region (centered at E_i with halfwidth Γ_i):

$$P(h\nu_1, h\nu_2, E_f) \propto \frac{2}{E_f - (h\nu_1 + h\nu_2) - E_i)^2 + \Gamma_i^2} \cdot \frac{2}{E_f - (h\nu_2) - E_j)^2 + \Gamma_j^2} \qquad (2.29)$$

The intermediate-state line halfwidth Γ_j implicitly contains relaxation due to inelastic scattering as well as the pure dephasing time[4]. Figure 2.3 indicates the relation of mea-

---

[4]in terms of a discrete level system, the electric field (due to photons) creates a coherent superposition of initial state, intermediate state and final state [113]. Quasi-elastic scattering events (due to defects, phonons; withdrawing tiny amounts of energy) lead to a decay of the phase of the electron (pure dephasing

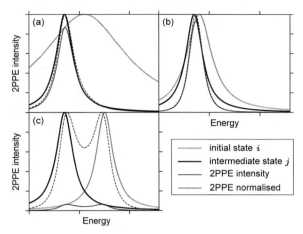

Figure 2.3: illustration of equation (2.29) for three cases: (a) $\Gamma_i \gg \Gamma_j$, (b) $\Gamma_i \approx \Gamma_j$ and $E_j = E_i + h\nu_1 + \varepsilon$, (c) $\Gamma_i = \Gamma_j$ and $E_j \neq E_i + h\nu_1$. The plotted curves for intermediate states are shifted in energy by $-h\nu_1$ and for 2PPE intensity by $-h\nu_1 - h\nu_2$.

sured linewidths (red) to the initial and intermediate state linewidths for three different constellations. In (a), the initial density of states (grey) is strongly broadened compared to the intermediate state. Then, the measured linewidth deviates only slightly from the intermediate state linewidth. In (b), both initial and intermediate states have similar (half) linewidths Γ_i, Γ_j and the measured half linewidth is considerably narrower than both Γ_i and Γ_j. The full width at half maximum of the measured 2PPE line shape in resonance is [110]

$$W_{E,\text{2PPE}} = 2\Gamma_i\Gamma_j/(\Gamma_i^2 + \Gamma_j^2)^{1/2} \qquad (2.30)$$

In (c), initial and intermediate state peaks are energetically separated. In that case both produce peaks in the 2PPE spectrum where the peak positions are slightly shifted with respect to $E_i + h\nu_1 + h\nu_2$ and $E_j + h\nu_2$. The absolute photoelectron yield may be low (see solid red curve).

Two implications are important: (1) Compared to the interplay of final state linewidths and initial state linewidths in one-photon photoemission, here we have a product of the Lorentzian lineshapes instead of a convolution, such that the combined effect is narrowing instead of broadening. (2) If the initial state density of states consists of broadened structures, the measured 2PPE linewidths reflect the linewidths of the intermediate state (case (a) in figure 2.3). The precondition can be independently checked by one-photon photoemission.

---

with a rate $1/T_2^*$). Inelastic scattering events lead to depopulation for each of the three levels with rates commonly denoted by the symbol $1/T_1$

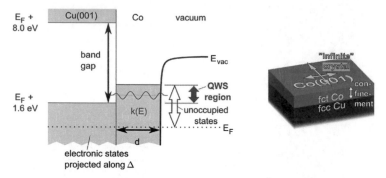

Figure 2.4: An example for an overlayer/substrate constellation (cobalt on fcc Cu(001)) that leads to formation of QWS in the overlayer. Gray regions indicate the existence of electronic states projected along the Δ direction in the fcc Brillouin zone.

2.7 Quantum well states in metallic thin films

Quantum well states (QWS) occur as a consequence of electron confinement in one, two or three dimensions. The simplest example of confinement in quantum mechanics is the particle in a box which has a discrete energy spectrum, whereas for the free electron, it is continous. An analogous situation is found for electrons in the periodic potential of a crystal.

The change of behaviour due to confinement is sometimes called quantum size effect. "Quantum size" is defined by the coherence length of the quantum object, i.e. the length scale over which the quantum mechanical phase is not destroyed (randomly changed) by external processes such as scattering. For a silver thin film on Au(111) Miller et al. [114] have deduced a lower limit of 100 Å for the coherence length of the observed quantum well states in silver. Furthermore, if the size is increased to macroscopic dimensions, the energy differences between discretized states decreases to orders of magnitude such that they may appear as continous or cannot be resolved experimentally: For a particle in a 1-dimensional 1-nm-wide box, the two lowest energy values are 1.13 eV apart whereas for a 100-nm-wide box, they are 0.113 meV apart.

In this work, one layer of Co on a Cu(001) substrate results in one-dimensional confinement. The formation of quantum well states for this system is shown in figure 2.4. In copper, electronic states with Bloch wave vectors $\vec{k} = (0, 0, k_z)$ along the Δ axis form a gap in the energy range from $E_F + 1.6$ eV to $E_F + 8$ eV [8, 29, 115]. For cobalt, the lower edge of this band gap is at $E_F + 3.0$ eV (minority spin) or $E_F + 2.8$ eV (majority spin). The energy regions where electronic states exist, are marked by the grey shaded areas in figure 2.4. Electrons in the cobalt layer that occupy states within the energy region indicated by the red arrow, are reflected at the copper substrate as well as at the surface barrier (to the vacuum side). Contrarily, in the surface parallel directions, the translation symmetry of the lattice is conserved and the crystal is macroscopically extended in these directions, such that the confined electrons form a 2-dimensional electron gas.

Mismatch in the relative position of band gaps occurs quite commonly, and, accordingly, quantum well states have been found for many thin film/substrate combinations, e.g. in in [116] combining ferromagnetic and non-magnetic transition metals. All of the examples in [116] have in common, that the quantum well states are derived from bulk states *close to the Brillouin zone boundary*. An example for occurence of quantum well states close to the Γ point is Mg/Si(111) [117].

In photoemission, resonances due to quantum well states can be observed to shift in energy when the film thickness is increased. Furthermore the density of quantum well states within a fixed energy interval increases. The phase accumulation model has been frequently used to interpret this energy dependence of quantum well states on film thickness [118, 119]. The wave function of the quantum well state is said to accumulate a phase of multiples of 2π during a round trip within the confined region of thickness d such that it forms a standing wave:

$$\phi_B(E) + \phi_C(E) + 2k_\perp(E)d = 2\pi n \qquad (2.31)$$

where ϕ_B, ϕ_C are phase jumps shown in figure 2.5. ϕ_B, ϕ_C can be calculated from a small number of parameters using empirical equations [119–122]:

$$\phi_B/\pi = \sqrt{\frac{3.4\,\mathrm{eV}}{E_V - E} - 1} \qquad (2.32)$$

$$\phi_C = 2\arcsin\left(\frac{E - E_L}{E_U - E_L}\right)^{1/2} - \pi \qquad (2.33)$$

where E_U, E_L are upper and lower limits of the relevant band gap of the substrate and E_V is the vacuum level. For $k(E)$, frequently a two-band nearly-free electron model is used [120]. Inherent in the phase accumulation model is the idea, that the wave function has to solve both the periodic potential of the ion cores (i.e. the bulk Hamiltonian) and the boundary conditions of the interfaces. The former is achieved by the rapidly varying Bloch wave, and the latter by the envelope. The physical reality of the envelope function has been demonstrated by experiments where quantum well states in Ag layers were coupled through a Au barrier layer (Au(111)+Ag+Au+Ag [123–125]). In summary, the phase accumulation model yields a simple description of the thickness dependence of quantum well state energies.

A number of experimental works have shown that photoemission experiments on quantum-well states can be used to determine the bulk band structure of materials [117, 126–128].

Quantum well states have been given much interest for their role in magnetic coupling across non-magnetic layers and magnetoresistance effects [116]. A typical feature of effects due to quantum well states is that physical quantities depend on the film thickness, and often so in an oscillatory manner. Oscillatory effects have been proven for the electric conductivity [129], for the Hall effect [130], for superconductivity [131] and magnetic anisotropy [132]. In case of the electric conductivity, the interpretation is that the energetic shift of quantum well states across the Fermi level E_F leads to an oscillating density of states at E_F. Furthermore, a thickness dependence of the magneto-optical Kerr effect

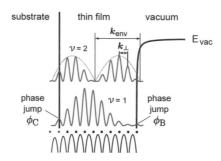

Figure 2.5: sketch of the probability density of two quantum well states (indices $\nu = 1, 2$) close to the Brillouin zone border. The periodic potential of the crystal lattice is indicated at the bottom; positions of the atoms are given by the black dots. According to the phase accumulation model the phase of the wave function changes within the quantum well according to $k_\perp$ and is subject to phase jumps during reflection at the thin-film/substrate interface (ϕ_C) and the vacuum barrier (ϕ_B). The grey dashed line is an envelope of the probability density and the number of its nodes defines the order ν of the quantum well state. The envelope function is discussed in more detail in [120].

was found for Co/Cu(001) [133]. Himpsel [134] reviewed the role of quantum well states in magnetic phenomena such as oscillatory coupling and giant magnetoresistance.

Chapter 3

Experimental methods

3.1 The momentum microscope

A schematic overview of the instrument is given in figure 3.1(a). The principle design of the momentum microscope without spin filter has been published in [13]. The momentum imaging technique is similar to that of a PEEM, where the photoelectrons are projected by electron lenses onto a position-sensitive detector. Projecting the focal plane rather than the image plane onto the detector leads to an intensity distribution which resolves the angles of photoemitted electrons. Figure 3.1(b) shows a sketch of the emitted photoelectron cone and the effect of the first electrode.

The initial emission angle α_0 is related to the parallel momentum component $\vec{p}_{||,0}$ by $\sin(\alpha_0) = |\vec{p}_{||,0}|/|\vec{p}_0|$. The anode strongly accelerates all emitted photoelectrons in the surface normal direction without changing the parallel-momentum component. Therefore,

$$\frac{\sin(\alpha_0)}{\sin(\alpha')} = \frac{|\vec{p}_{||,0}|/|\vec{p}_0|}{|\vec{p}_{||}'|/|\vec{p}'|} = \frac{|\vec{p}_{||,0}|/|\vec{p}_0|}{|\vec{p}_{||,0}|/|\vec{p}'|} = \frac{|\vec{p}'|}{|\vec{p}_0|} \tag{3.1}$$

$$= \sqrt{1 + \frac{eU_{\text{ext}}}{E_{\text{kin}}}} \approx \sqrt{\frac{eU_{\text{ext}}}{E_{\text{kin}}}} \tag{3.2}$$

$$\sin(\alpha') = \sin(\alpha_0)\sqrt{\frac{E_{\text{kin}}}{eU_{\text{ext}}}} \ll 1 \tag{3.3}$$

$$\sin(\alpha') = \frac{|\vec{p}_{||,0}|}{\sqrt{eU_{\text{ext}}}\sqrt{2m}} \tag{3.4}$$

Between equations (3.3) and (3.4), we have replaced $\sin(\alpha_0) = |\vec{p}_{||,0}|/\sqrt{2mE_{\text{kin}}}$. The accelerating voltage eU_{ext} is chosen large compared to the photoelectron kinetic energy E_{kin}. Hence, the emission cone entering the electron optics is quite narrow (α' small - see equation (3.3)). The lateral position of a photoelectron beam within the focal plane is proportional to $\sin(\alpha')$ and by equation (3.4) also proportional to the parallel-momentum component $\vec{p}_{||,0}$ of the photoelectron. Therefore, when the focal plane is projected onto a detector, we obtain an intensity distribution of the *parallel momenta* of the photoelectrons.

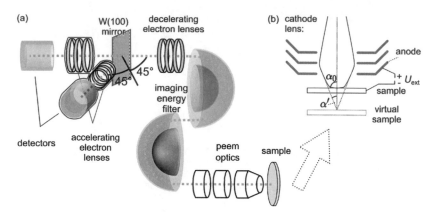

Figure 3.1: (a) Momentum microscope with energy analyzer and spin polarization analyzer designed as 2-dimensional imaging devices, that allow highly efficient parallelised
measurements of spin-resolved photoemission spectra. The parts outlined in red form
the spin-resolved detection branch with the inhouse designed spin polarization analyser.
In this work, these were attached to a commercially available instrument (Omicron NanoESCA [13, 135] — parts outside the red outline). (b) The ensemble of photoelectrons
emitted into the half-space ($0 < \alpha_0 < 90°$) above the sample forms a cone with small angle
α' after being strongly accelerated by the anode. This leads to an image in the focal plane
where the lateral position is proportional to the parallel-momentum component rather
than the emission angle (see text).

The instrument used in this work enables both spatial imaging mode and momentum
imaging mode through the use of a transfer lens that allows to shift the depth position of
focal or image planes such that either one of them can be placed onto the entrance slit of
the energy analyzer. The geometry of the objective lens part enables a field of view in the
diameter range from 20 to 80 μm in the spatial imaging mode and 0.5 to 2.5 Å^{-1} in the
momentum imaging mode.

An important aspect contributing to the excellent momentum resolution is the double hemispherical analyzer. The first hemisphere yields an energy-filtered image at its
exit while the second hemisphere reverses the effect of aberrations introduced by the first
hemisphere. Since the electrons move in an $1/r$ potential present between the inner and
outer shells of each hemisphere, their trajectory would be a closed ellipse – the only difference is that the second hemisphere is rotated 180° around the optical axis going through
its entrance. This means, in the exit plane of the second hemisphere, the positions and
angles of all electron beams are the same as in the entrance plane of the first hemisphere.
The energy resolution (and transmission) can be set by adapting the pass energy (kinetic
energy of the photoelectrons while passing through the hemispheres) or by choosing an
aperture size at the entrance (and exit) of the energy analyzer. In our experiments the
typical energy resolution was 0.1 eV.

Figure 3.2 shows an energy diagram of the photoemission process and the energy
filter. The final state energy is selected by tuning the sample voltage. The sample voltage

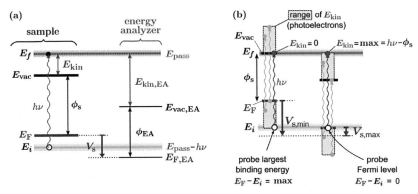

Figure 3.2: (a) Scheme of the photoemission process at the sample and the energy selection at the energy analyzer. V_s – the potential difference between the Fermi levels of sample and entrance slit of the energy analyzer – is used to tune the selected final state energy E_f (and simultaneously the initial state energy $E_i = E_f - h\nu$). Since V_s defines the relative positions of E_F, but the energy filtering is sensitive to E_{kin}, the work functions of sample ϕ_s and energy analyzer ϕ_{EA} have to be taken into account. The green horizontal bar at E_{pass} indicates, which photoelectrons pass through the energy filter; the second horizontal bar at $E_{pass} - h\nu$ corresponds to the initial state energies. (b) Tuning the sample voltage: Left and right hand side show the situation for two limiting sample voltages where the slowest (left, $E_{kin} = 0$) or fastest (right, $E_{kin} = h\nu - \phi_s$) photoelectrons are probed. Beyond the marked energy range, the photoelectron current drops to zero. In figure (b), $V_{s,min} = -(E_{pass} - E_{F,EA} - \phi_s)$ and $V_{s,max} = -(E_{pass} - h\nu - E_{F,EA})$ From the difference of these two sample voltages, one obtains $V_{s,max} - V_{s,min} = h\nu - \phi_s$, which yields the work function of the sample without requiring $\phi_E A$.

determines the relative position of the Fermi level of sample and energy analyzer, thereby reducing or enhancing the kinetic energy of photoelectrons with respect to the pass energy.

3.2 Electron spin detection principles

The most commonly applied way to analyze the photoelectron spin is Mott scattering [136, 137]. Photoelectrons, whose spin polarization shall be determined, are accelerated to energies $30 \ldots 100$ keV and scattered at a thin gold foil with normal incidence. Elements with a high atomic number Z are best suited (gold: Z=79), since high electric field strength in the vicinity of the atom cores enhance the effect of spin-orbit coupling. Thin foils are used to diminish the occurence of multiple scattering. Count rates of backscattered electrons are measured at an angle θ with detectors to the left I_l and to the right I_r. θ is chosen such that the Sherman function $S(\theta)$ is maximum. $S(\theta)$ relates the left/right count rate asymmetry to the spin polarization component $\vec{P} \cdot \hat{n}$ along the quantization axis $\hat{n}$:

$$\frac{I_l - I_r}{I_l + I_r} = S(\theta)P \tag{3.5}$$

where $\hat{n}$ is the unit vector perpendicular to the scattering plane.

In order to minimize the statistical error of a spin-polarized measurement, it can be shown, that the product S^2N should be as large as possible, where N is the number of detected particles [86]. N scales proportionally to the absolute scattering cross section (here referred to as reflectivity R), so that S^2R should be maximised. This number is defined as the figure of merit. It enables a comparison of the efficiency of spin polarization analyzers. A value of 1 means that, in case of the single-channel Mott detector, every electron is reflected and a fully polarized electron beam will be scattered exclusively to the left (or the right) side. Typical figures of merit for Mott detectors are in the order of 10^{-4}, with a reflectivity of ≈ 0.01 and the (effective) Sherman function 0.16 [138].

A further type of spin-polarization analyzers is based on low-energy electron diffraction (LEED). The spin-dependent mechanism is again based on either spin-orbit coupling or exchange interaction. Here, multiple scattering of the impinging electron is a desired effect and enables scattering conditions where high reflectivity may occur simultaneously with high sensitivity on the electron spin [139], while for scattering electrons at atoms, high spin sensitivity is generally paired with low reflectivity. As a side effect of the low energies, the spin-dependent scattering mechanism is much more sensitive to the cleanliness and ordering of the surface and therefore requires a careful preparation of the crystal surface and ultra-high vacuum conditions.

Spin analyzers based on exchange interaction have also been used for photoemission spectroscopy [140, 141]. In this case, the working principle is most easily understood by considering the exchange-split majority and minority electron bands. The position of band gaps then also depends on the spin. In low-energy electron diffraction, the vanishing density of states within the band gaps corresponds to high reflectivity. Therefore, electron scattering on magnetic surfaces with low energy is supposed to feature high reflectivity as well as high spin asymmetry. The magnetisation of the spin analyzer can be oriented to select the spin polarization component which is analyzed. Two oppositely magnetised measurements are used to calculate an intensity asymmetry, which leads to the spin polar-ization. In [141], Fe/W(001) with chemisorbed $p(1 \times 1)$ oxygen was used, and a Sherman function of 0.24($\pm$0.03), a reflectivity of 0.076, and a figure of merit $\mathcal{F} \approx 2.2 \times 10^{-3}$ were achieved for an electron scattering energy 13.5 eV.

3.3   Spin-resolved photoemission measurements

The spin-resolved photoemission measurements in this work were performed using a novel spin filter principle, which achieves a significant reduction in measurement time due to parallelisation. When spin-filtering an image instead of a single data point, the number $\mathcal{N}$ of resolvable image points is equivalent to using $\mathcal{N}$ single-channel spin detectors in parallel, so that the efficiency of the imaging spin detector is increased by $\mathcal{N}$. As stated in [14], this number is limited by the angular resolution of the electron diffraction in the W(100) surface. From a spin-filtered image of a magnetic domain wall, we found, that 70 discrete points were resolved along the diameter of the field of view, resulting in $\mathcal{N} = \pi * 70^2 \approx 3800$ for the 2D-image. In addition to the efficiency increase due to the imaging principle, the value of the figure of merit $S^2R = 0.42^2 * 0.012 = 2.1 \times 10^{-3}$ in terms of single-channel spin filters competes well with other reported values [141, 142] (the upper limit being at

$\approx 1 \times 10^{-2}$ based on very-low energy electron diffraction while Mott type spin polarization analyzers typically achieve $\approx 1 \times 10^{-4}$).

The position of the spin filter within the momentum microscope is given in figure 3.1(a). In the exit slit of the energy analyzer, a reciprocal image of either the spatial or momentum image is formed. A lens system projects this reciprocal image onto the W(100) surface. Figure 3.3(a) shows a close-up of the W(100) surface and the electron beams that transport the image information. Every position of the image is encoded to a different scattering angle. The W(001) crystal is oriented such that the central image point is incident at a polar angle of $45°$ and along an azimuthal angle $0°$ corresponding to the in-plane (10) direction of the bulk crystal[1]. The off-center image points correspond to angular deviations within a range of typically $\pm 1.5°$ as obtained by electron-optical simulations [144]. For the photoelectrons which are scattered into the (00) LEED spot, the angles are transformed the same way as for light rays reflected by a mirror. Hence, the geometrical arrangement of the individual intensity channels forming the image is conserved. Contrarily, the intensity itself is not conserved, since the scattering amplitude of photoelectrons at the W(100) surface is spin-dependent. An example is shown in figure 3.3(b): There, the probability for scattering a spin-down electron is lower than for a spin-up electron. In this example, if we scattered 100 spin-up electrons at the W(100) surface, we would count a higher number of electrons after the spin filter than if we scattered 100 spin-down electrons. This means, the incident intensity is equal in both cases, but the intensity after scattering at the W(100) surface depends on the spin polarization of the electron beam.

One further aspect of the imaging spin filter is important: For a Mott detector, spin-polarized electrons are scattered to the left and right. Two intensities are being detected, which allow to calculate the spin polarization as well as the spin-integral (total) intensity. As can be seen in figure 3.1(a), we detect electrons scattered to the left, but not to the right, as this would require two detection arms as well as constantly flipping the W(100) crystal around 90° back and forth. This is clearly not desirable as it would cause a lot of problems (mechanical instability, friction processes that deteriorate the ultra-high vacuum, instrumental asymmetries of the two detection arms, etc.) However, if we measure only a single scattered intensity value, it is not possible, to calculate two independent quantities (total intensity I_0 and spin polarization P). Two different approaches to solve this problem, are shown in this work: In spatial imaging mode, we select a spot on the sample where a magnetic domain wall separates two areas that are magnetized anti-parallel to each other. This yields two intensity values, from which I_0 and P can be calculated. We used this approach for the characterization of the spin detector (section 3.3.1) The second – more general – approach is, to tune the spin sensitivity of the scattering process by changing the kinetic energy of photoelectrons when they scatter at the W(100) surface. Again, one can record two independent intensity values that both depend on the spin polarization and incident intensity. This will be discussed in detail in section 3.3.2.

The configuration of electron lenses used to project the reciprocal image onto the W(100) crystal allows for adjustment of the kinetic energy from $\approx 10\,\mathrm{eV}$ to more than

---

[1](Azimuthal) rotation diagrams of the reflectivity and spin polarization in LEED from W(100) have been measured and calculated in [143]. The primary energy was 100 eV. The highest experimental spin polarization occured in the in-plane (100) direction, paired with a sizeable reflectivity well above the minimum value.

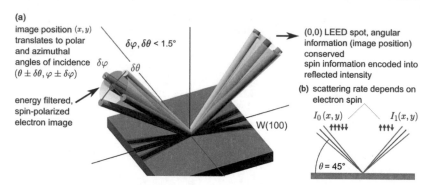

Figure 3.3: working principle of the imaging spin filter, scattering geometry at the W(100) surface; the photoelectrons are scattered such that only their momentum component perpendicular to the W(100) surface changes. For distinctness, some of the beams have been given individual colors.

100 eV (the precise range depends on the pass energy) without changing the magnification of the projected image.

While the influence of the *scattering angle* is a mostly smoothly varying function that shall be eliminated, the dependence on the spin and on the *scattering energy* is exploited, to obtain the spin polarization, or more precisely – its projection onto the quantization axis).

The spin-dependence of the scattering amplitude is caused by spin-orbit coupling in the W(001) crystal. Therefore, this spin quantization axis is perpendicular to the scattering plane.

3.3.1   Characterization of the imaging spin polarization analyzer

Figure 3.4 shows a two-photon photoemission measurement with p-polarized light in spatial energy-resolved microscopy mode on an 8 monolayer Co/Cu(001) film. In the back-focal plane of the objective lens, an aperture of 70 μm limits the parallel-momentum range of photoelectrons contributing to the spatial image to about 0.1 Å^{-1} centered around normal photoemission. The spin-polarization of photoelectrons in two-photon photoemission using 3.0 eV p-polarized photons on Co/Cu(001) thin films has been shown to be within $(45 \pm 5)\%$ for most of the accessible binding energy range [21]. In our case, a constant initial state energy of $E_F - 1.2$ eV was set (resolution of the energy analyzer was 0.1 eV), where the spin-polarization slightly drops to about 40%. For the various datapoints in figure 3.4(a), the scattering energy of the photoelectrons at the W(001) surface has been varied, and an intensity value has been recorded by averaging over the areas marked in the spatial images in figure 3.4(b). Throughout the experiment, the sample was kept in the same condition. That means, the same intensity distribution of photoelectrons was emitted from the sample for every datapoint. Then, the recorded change in the intensities happens only after scattering from the W(001) crystal and the intensities from each

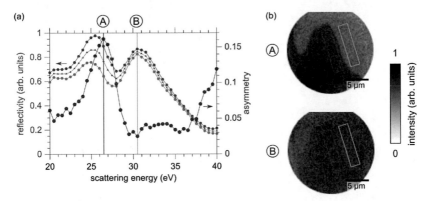

Figure 3.4: (a) scattering energy dependence at the W(001) crystal of the intensity measured from two magnetic domains in spatial mode (blue points), their average (grey triangles) and their asymmetry (red points) in two-photon photoemission on an 8 monolayer Co thin film on Cu(001) (lines are guides to the eye); (b) the spatial images with rectangles showing the area, where the intensity was integrated to yield the datapoints in (a). The exposure time was 60 seconds.

magnetic domain (indices $+, -$) are proportional to

$$I_{\pm}(E_{\text{scatt}}) = I_0(1 + S(E_{\text{scatt}})P_{\pm})R(E_{\text{scatt}}) \tag{3.6}$$

As shown in figure 3.1(a), the W(001) can be retracted from the electron-optical path to detect the photoelectrons without spin-filtering. In that case, the spatial image shows a constant intensity across the whole field of view. Therefore, the intensity contrast observed in the spatial images on the right hand side of figure 3.4 is caused by different spin polarization emitted from different magnetic domains. These were present on the Co film after growth. Azimuthal rotation of the sample by angles of $10°$ has been verified to lead to decreased contrast, which vanishes for the $90°$-rotated case. This tells us, that the two areas are magnetized along the spin quantization axis (= vertical axis in the image) with opposite sign of the magnetic dipole. If we assume a negligible influence of spin-orbit coupling, the spin polarization of each domain has the same absolute value and opposite sign. Inserting $P := P_+ = -P_-$ in equation (3.6), we obtain $S(E_{\text{scatt}})$ and $R(E_{\text{scatt}})$ as follows:

$$I_+ + I_- = 2R \cdot I_0 \tag{3.7}$$

$$\frac{I_+ - I_-}{I_+ + I_-} = S \cdot P \tag{3.8}$$

In non-magnetic materials, the spin polarization of photoelectrons is solely generated by spin-orbit coupling. One particular effect has been theoretically [145] predicted and experimentally [146] confirmed for (001) surfaces of non-magnetic fcc crystals. Off-normally incident p-polarized light produces a spin polarization component perpendicular to the

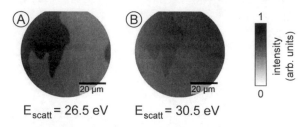

Figure 3.5: spatial images with exposure time 120 s and larger field of view demonstrating more clearly the effect of varying spin sensitivity at the two scattering energies $E_{\text{scatt}} = 26.5\,\text{eV},\ 30.5\,\text{eV}$

plane of incidence (spanned by the surface normal and the light propagation vector) for normally emitted photoelectrons. In our photoemission geometry, this spin polarization component coincides with our spin-sensitive axis. From experiments, the degree of spin polarization has been shown to be below 15% using 21.2 eV and 16.9 eV unpolarized photon beams on a Pt(001) surface [146]. Platinum being one of the heavy elements, we expect the effect to be less strong for cobalt (atomic numbers $Z_{\text{Pt}} = 78$, $Z_{\text{Co}} = 27$). In section 3.3.3, the calculated values of P for copper (Z=29) are below 4 %. The error introduced by neglecting the spin-orbit coupling, introduces an error of few percent to the values of the spin sensitivity $S(E_{\text{scatt}})$ as obtained from equation (3.8).

In figure 3.4(a), two scattering energies have been marked with (A) and (B), where the spin sensitivity reaches a local maximum for (A) and a local minimum for (B). In both cases, the average reflectivity (shown as gray curve) is comparable and close to the maximum value observed for the measured scattering energy range. If we use the spin polarization $P = 40\%$ for an initial state energy of $E_{\text{F}} - 1.2\,\text{eV}$, the maximum spin sensitivity $S_{26.5\text{eV}}$ follows from dividing the asymmetry value 0.16 by P which yields $S_{26.5\text{eV}} = 0.4 \pm 0.05$. The spin sensitivity for all other scattering energies is obtained likewise. At scattering energies above 30.5 eV, the intensity drops to one fifth of the maximum which is unfavourable in terms of efficiency.

The effect of switching the scattering energy from 26.5 eV to 30.5 eV can be seen more clearly in figure 3.5. The contrast does not vanish in the case (B) but is much weaker than for (A). These two scattering energies are interesting working points, because they allow to distinguish the effect of the electron spin even when a direct comparison of simultaneously measured photoelectrons from two magnetic domains is not possible. This is an important experimental means that enables the spin-resolved measurement of *momentum images*.

3.3.2 Evaluation of the spin polarization

In this section, a method is given which allows to obtain the spin polarization component P along the quantization axis for parallel-momentum distributions of photoelectrons. As noted before, this is achieved measuring the scattered intensities I_l, I_h for two different scattering energies (indexed by l, h) and carrying out an analysis, that yields P and the spin-integrated ("total") photocurrent I_0.

The problem of elastic scattering of an electron from an atom is treated in a textbook by Kessler [86], which leads to the following expression for the differential cross section σ

$$\sigma(\theta,\phi) = R(\theta)\left(1 + S(\theta)\,\vec{P}\cdot\hat{n}\right) \qquad (3.9)$$

where θ,ϕ are the polar and azimuthal angles, $\hat{n}$ is the scattering plane normal, $R(\theta)$ is the spin integrated scattering amplitude and $S(\theta)$ is the spin sensitivity, also denoted as Sherman function. In a solid, the potential is not spherically symmetric and also due to multiple scattering, scattering amplitude and spin sensitivity depend as well on the azimuthal angle ($R = R(\theta,\phi)$ and $S = S(\theta,\phi)$) [144]. In the following, we replace the angles by the image position (x,y) to which they correspond and denote the spin component perpendicular to the scattering plane by $P = \vec{P}\cdot\hat{n}$. Since the scattering cross section is proportional to the ratio of scattered photo current I_l, I_h to incoming photo current I_0, one can write for the scattered intensities obtained by measuring at two different scattering energies:

$$I_h = I_0(x,y)\left(1 + S_h(x,y)\,P(x,y)\right) R_h(x,y) \qquad (3.10)$$

$$I_l = I_0(x,y)\left(1 + S_l(x,y)\,P(x,y)\right) R_l(x,y) \qquad (3.11)$$

The spin-integrated photocurrent I_0 can be eliminated by dividing eq. (3.10) by (3.11). Additionally, division by R_h and R_l leads to

$$Y(x,y) = \frac{1 + S_h(x,y)\,P(x,y)}{1 + S_l(x,y)\,P(x,y)}, \qquad (3.12)$$

where $Y(x,y) = \frac{R_l(x,y)}{R_h(x,y)}\frac{I_h(x,y)}{I_l(x,y)}$. This can be solved for P – the spin polarization component perpendicular to the scattering plane

$$P(x,y) = \frac{Y(x,y) - 1}{S_h(x,y) - Y(x,y)\,S_l(x,y)} \qquad (3.13)$$

From the introduced abbreviation $Y(x,y)$ it is obvious, that the spin polarization depends only on the ratio of the spin-averaged scattering amplitudes R_l, R_h and the ratio of scattered intensities I_h, I_l. In order to derive the spin-integrated incident photocurrent I_0, equation (3.13) can be used to eliminate $P(x,y)$ in equation (3.10). After algebraic transformations the expression can be simplified to

$$I_0(x,y) = \frac{I_h(x,y)S_l(x,y)/R_h(x,y) - I_l(x,y)S_h(x,y)/R_l(x,y)}{S_l(x,y) - S_h(x,y)} \qquad (3.14)$$

Here, the absolute values of the spin-integrated reflectivities $R_l(x,y), R_h(x,y)$ occur separately. These can be obtained by projecting an unpolarized ($P = 0$) electron image with uniform intensity ($I_0(x,y) = i_0$) onto the W(100) surface. The right-hand-side of equation (3.10) then reduces to $i_0\,R(x,y)$ and measuring the unpolarized homogeneous image for both scattering energies yields $R_l(x,y), R_h(x,y)$ with the same multiplicative constant i_0. The generation of an unpolarized electron image is discussed section 3.3.3. Figure 3.6 shows the ratio of the spin-averaged scattering amplitudes R_l and R_h resolved in polar

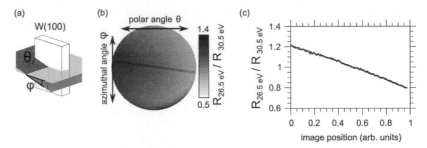

Figure 3.6: (a) sketch showing the polar angle θ and the azimuth ϕ for a scattered electron beam (red line) (b) ratio of intensities measured with unpolarized electrons at two scattering energies in dependence of the polar and azimuthal angles (c) line profile along the blue line in (b); the amount of variation of $R_{26.5\,eV}\,/\,R_{30.5\,eV}$ within the field of view grows with the range of polar and azimuthal angles that is necessary to transmit the reciprocal energy-filtered image. The data shown in (b,c) was typical for the measurements performed in our home laboratory.

and azimuthal angles for the scattering energies $E_{scatt,h} = 26.5$ eV and $E_{scatt,l} = 30.5$ eV. As shown by the definition of Y, this ratio can be directly used to eliminate the angular dependence of the scattering amplitudes from the measured intensities. The settings of the electron lenses between the tungsten crystal and the detector determine the mapping between scattering angles and image position. These have to be the same for obtaining R_l, R_h and for performing the actual spin-resolved photoemission experiment.

When P and I_0 have been calculated, an alternative representation in terms of spin-up and spin-down partial intensities $I_\uparrow, I_\downarrow$ is easily obtained as follows:

$$I_\uparrow(x,y) \;=\; \frac{I_0(x,y)}{2}\left(1 + P(x,y)\right) \tag{3.15a}$$

$$I_\downarrow(x,y) \;=\; \frac{I_0(x,y)}{2}\left(1 - P(x,y)\right) \tag{3.15b}$$

3.3.3 Generation of an unpolarized electron image from Cu(001)

As pointed out in the previous section, we need to obtain the (angular resolved) reflectivities $R_h(x,y), R_l(x,y)$ for scattering at the W(100) surface with kinetic energies $E_h = 26.5$ eV and $E_l = 30.5$ eV. (The image position (x,y) translates to polar and azimuthal scattering angles). Therefore, an unpolarized, electron image with homogeneous intensity is required. We have used the defocussed spatial electron image from a Cu(001) surface which is illuminated with unpolarized light from a mercury arc lamp. Even when recording a spatial image, an aperture sitting in the back focal plane of the sample selects a subset of the parallel-momentum $(k_{||})$ range of photoelectrons. For the generation of the unpolarized electron image, we opened this aperture, such that the whole parallel-momentum distribution of photoelectrons contributes to the defocussed spatial image. The requirement of vanishing average spin polarization was checked by performing a calculation of the $k_{||}$-resolved photoemission from a Cu(001) surface with an unpolarized photon beam ($h\nu = 4.9\,\mathrm{eV}$) [147]. Typical spectra of mercury arc lamps show an intensity peak at the high-energy cutoff of the spectrum centered around 4.88 eV [148]. Assuming a work function of 4.5 eV for copper [149], the energy distribution of photoelectrons ranges up to $E_{\mathrm{kin}} = 0.4\,\mathrm{eV}$. Figure 3.7 presents the calculated momentum distributions of the spin-polarized photocurrent. The relative orientation of the direction where the mercury arc lamp points and the scattering plane of the spin detector determines whether the $k_\parallel$-averaged distribution of photoelectrons is spin polarized or not. In our home lab, the optical plane (defined by incidence of the photons and surface normal) was perpendicular to the spin-sensitive axis (case (b)), while for measurements at ELETTRA, Trieste, the optical plane was parallel to the spin-sensitive axis (case (a)). The conclusions for generating an ideally unpolarized photoelectron beam necessary for the measurement of $R_l(x,y), R_h(x,y)$ are: Illumination of a Cu(001) surface with a mercury arc lamp is suitable for generating an electron image with nearly vanishing spin polarization. Equal azimuthal angles for photon propagation and spin-sensitive axis leads to a $k_\parallel$-averaged spin polarization of 0, if the selected $k_\parallel$-range is centered around normal emission (figure 3.7(a)). A relative azimuthal angle of 90° between photon propagation and spin-sensitive axis yields a $k_\parallel$-averaged spin polarization below 3 %(figure 3.7(b)). Selecting a kinetic energy of 0.4 eV, which yields the highest photocurrent, reduces this value to 1.2 %.

3.3.4   Analysis of the propagation of experimental errors

Since the spin polarization depends on the variable $Y(x,y) = \frac{R_l(x,y)}{R_h(x,y)} \frac{I_h(x,y)}{I_l(x,y)}$, it is possible to derive in one step the error $u(P)$ of the spin polarization with respect to an error of the measured intensity ratio $r_I = I_h/I_l$ as well as to the reflectivity ratio $r_R = R_l/R_h$:

$$u(P) = \sqrt{\left(\frac{\partial P}{\partial r_I}\right)^2 u^2(r_I) + \left(\frac{\partial P}{\partial r_R}\right)^2 u^2(r_R)} \tag{3.16}$$

$$u(P) = \frac{\partial P}{\partial Y}\left[\left(\frac{\partial Y}{\partial r_I}\right)^2 u^2(r_I) + \left(\frac{\partial Y}{\partial r_R}\right)^2 u^2(r_R)\right]^{1/2} \tag{3.17}$$

$$u(P) = \frac{S_h - S_l}{(S_h - Y S_l)^2}\left[\left(\frac{R_l}{R_h}\right)^2 u^2\left(\frac{I_h}{I_l}\right) + \left(\frac{I_h}{I_l}\right)^2 u^2\left(\frac{R_l}{R_h}\right)\right]^{1/2} \tag{3.18}$$

When approximating $Y, \frac{I_h}{I_l}, \frac{R_l}{R_h} \approx 1$ (which for our experiments with W(100) is a very good approximation for $R_{l,h}$ and in case of low spin polarizations holds for $I_{l,h}$ as well), we arrive at

$$u(P) \approx \frac{1}{S_h - S_l}\left[\frac{R_l}{R_h} u\left(\frac{I_h}{I_l}\right) + \frac{I_h}{I_l} u\left(\frac{R_l}{R_h}\right)\right], \tag{3.19}$$

The error of the calculated spin polarization gets smaller with increasing difference $S_h - S_l$, i.e. selecting two scattering energies such that their spin sensitivity differs as much as possible (even with a sign change if available) improves the efficiency of spin-resolved measurements.

A short numerical example illustrates the electron counts necessary to decrease the error of the calculated spin polarization to an acceptable value: The number of electrons

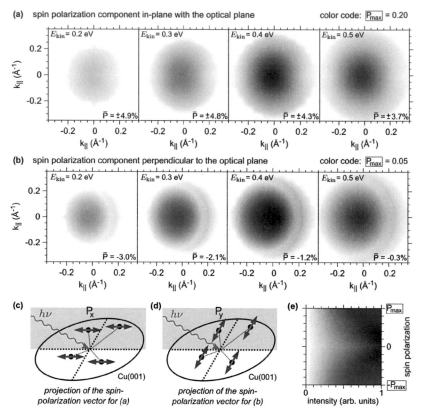

Figure 3.7: photoemission calculation using OMNI [147]. A Cu(001) surface illuminated by unpolarized light, $h\nu = 4.9\,\text{eV}$. The geometries for (a,b) are shown in (c,d). P_{max} denotes the spin polarization range for the color code in (e). Averages of the spin polarization $\bar{P}$ weighted with the intensity are given in the graphs. For (a), all $\bar{P}$ values are zero, the average values given in the graphs refer to the upper and lower halves of the momentum distributions.

N reaching the detector in a given time interval is Poisson distributed [150]. The standard deviation of this distribution depends on the mean value $\bar{N}$ as follows: $\sigma_N = \sqrt{\bar{N}}$. The relative error of the photoelectron count in case of 1000 recorded events is 3.2%. The absolute error $u(I_h/I_l) = (u(I_h)/I_h + u(I_l)/I_l) I_h/I_l$ is twice as large since it sums up two count numbers. Furthermore, if we set $u(R_{l,h}) = 0$, $S_h = 0.42$, $S_l = 0.05$, we obtain an absolute error of $u(P) = 18\%$ for the spin polarization of one image point, where 1000 events have been registered. For 10000 events, the same calculation yields $u(P) = 5.7\%$.

In order to compare the efficiency of our spin filter in combination with the presented measurement and data analysis scheme, the figure of merit $\mathcal{F}$ needs to be slightly adaptated. For Mott detectors, the error of the spin polarization can be expressed

$$u(P)_{\text{Mott}} = \sqrt{\frac{1}{S_{\text{eff}}^2 N}} \tag{3.20}$$

where N is the number of detected electrons. It can be deduced by calculating how the error due to counting statistics enters into the error of the intensity asymmetry [86]. If we follow the same approach, we have to focus on the term $u\left(\frac{I_h}{I_l}\right)$ in equation (3.19) and set $u(R_l/R_h)$ to zero, treating R_h/R_l as well as S_h, S_l as known parameters. The error of the intensity ratio $u(I_h/I_l)$ can be shown to be approximately proportional to $1/\sqrt{I_l}$:

$$u^2\left(\frac{I_h}{I_l}\right) = \left(\frac{1}{I_l}\right)^2 u^2(I_h) + \left(-\frac{I_h}{I_l^2}\right)^2 u^2(I_l) \tag{3.21}$$

$$= \left(\frac{1}{I_l}\right)^2 I_h + \left(-\frac{I_h}{I_l^2}\right)^2 I_l \tag{3.22}$$

$$= \frac{1}{I_l}\left(\frac{I_h}{I_l} + \left(\frac{I_h}{I_l}\right)^2\right) \approx \frac{2}{I_l} \tag{3.23}$$

If we simplify the problem again to an average reflectivity (setting both R_h, R_l to R_{avg}), then I_h, $I_l = \frac{R_{\text{avg}}}{2} I_0$ results in $u^2\left(\frac{I_h}{I_l}\right) \approx 4I_0/R_{\text{avg}}$. Inserting this into equation (3.19), we find

$$u(P) \approx \sqrt{\frac{4I_0}{(S_h - S_l)^2 R_{\text{avg}}}} \tag{3.24}$$

The figure of merit for this type of spin measurement can thus be written

$$\mathcal{F} = (S_h - S_l)^2 R_{\text{avg}}/4 \tag{3.25}$$

For an ideal spin-analyzing crystal, the term $S_h - S_l$ evaluates to 2, $R_{\text{avg}} = 0.5$, and $\mathcal{F} = 0.5$, rather than 1. The ideal value being lower than 1 can be interpreted as follows: Mott-type detectors record two diffracted beams ("left" and "right"), simultaneously. For our spin detector, electrons are scattered into the single (00) beam. Thus for the *ideal Mott* spin detector crystal, an unpolarized electron beam is split into halves of the primary intensities and a fully polarized beam is scattered fully to the right *or* left. In both cases, all of the primary electrons are counted in one of the detection channels. Contrarily, for low-energy electron diffraction in the (00) beam, it's impossible to reflect all electrons and realize 100%

spin asymmetry, at the same time: If we assume 100% spin asymmetry, an unpolarized beam can be reflected with maximum 50% of the primary intensity (e.g. every spin-up electron and no spin-down electron). Hence, the maximum possible reflectivity depends on the spin asymmetry. For the characterization of the W(001) surface as a spin filter, we have measured the reflectivities using an unpolarized primary electron beam. Hence, this effect is already included in the experimental R_{avg} value. Inserting $S_h - S_l = 0.37$ and $R_{\mathrm{avg}} = 0.012$ into equation (3.25), we obtain $\mathcal{F} = 4 \times 10^{-4}$. As mentioned at the beginning of section 3.3, we analyse the spin of an electron image consisting of 3800 intensity channels, in parallel. This increases our efficiency to $\mathcal{F}_{\mathrm{2D}} = 1.56$, i.e. our spin detector is 1.56 times more efficient than the hypothetical, ideal single-channel Mott detector.

Systematic errors P, I_0 by the uncertainties in $S_{h,l}, R_{h,l}$

During the evaluation of P, I_0 by equations (3.13) and (3.14), remaining uncertainties in $S_{h,l}, R_{h,l}$ result in deviations of the calculated values $P^{(u)}$, $I_0^{(u)}$ from the true values $P^{(0)}, I_0^{(0)}$. Here, we present a numerical evaluation of these deviations, which require no approximations. The following procedure is performed: (1) Assume electron beams with $I_0^{(0)} = 1$ and $P^{(0)} \in \{0; +1; -1\}$. (2) Calculate the scattered intensities (by equations (3.10), (3.11)) using the actual values for the W(100) surface in a particular state (that depends on the preparation history): $S_h^{(0)} = 0.42$; $S_l^{(0)} = 0.05$; $R_h^{(0)} = R_l^{(0)} = 1.0$. (3) Calculate $P^{(u)}$ and $I_0^{(u)}$ from the result of step 2 using $S_{l,h}^{(u)} = S_{l,h}^{(0)} + u(S_{l,h})$; $R_{l,h}^{(u)} = R_{l,h}^{(0)} + u(R_{l,h})$. (4) Evaluate the errors $u(P) = P^{(u)} - P^{(0)}$ and $u(I_0) = I_0^{(u)} - I_0^{(0)}$. The results are specific to the values of $S_{h,l}^{(0)}, R_{h,l}^{(0)}$ and hence are valid for the W(100) surface in combination with the particular scattering geometry and the selected scattering energies, which were used in this work. They are presented in figure 3.8.

From figure 3.8(a), we see that over- or underestimating the difference in spin sensitivities $S_h - S_l$ has no effect on unpolarized photo currents. For spin-polarized photo currents, too small values of $S_h - S_l$ drive the resultant value for the spin polarization $P^{(u)}$ towards $\pm\infty$ conserving the true sign of $P^{(0)}$. On the other hand, too high values of $S_h - S_l$ drive the spin polarization towards zero. The consequences are

(1) Assume, that $|P^{(u)}|$ obtained from the analysis is lower than the actual value $|P^{(0)}|$. By definition of P, the asymmetry between the spin-up and spin-down partial intensities $I_\uparrow^{(u)}, I_\downarrow^{(u)}$ obtained from analysis is then too low, i.e. the separation of photocurrent into spin-up and spin-down channel is decreased. For example, a photoemission peak that actually is present in only one of the spin channels, is then found to appear in both the $I_\uparrow^{(u)}$ spectrum and the $I_\downarrow^{(u)}$ spectrum[2].

(2) Assume, that $|P^{(u)}|$ obtained from the analysis is higher than the actual value $|P^{(0)}|$. Then, a photoemission peak that actually is present in only one of the spin channels results in partial intensity spectra $I_\uparrow^{(u)}, I_\downarrow^{(u)}$, where the peak has too much amplitude in one spin-channel and negative amplitude in the other spin-channel.

---

[2] $S_h - S_l$ may be underestimated, if, for S_h, the value of a freshly prepared W(100) surface is used although the spin-filtered photoemission experiment is performed for several hours without any refreshing of the W(100) surface in between. The actual spin sensitivity S_h gradually decreases as will be shown section 3.3.5.

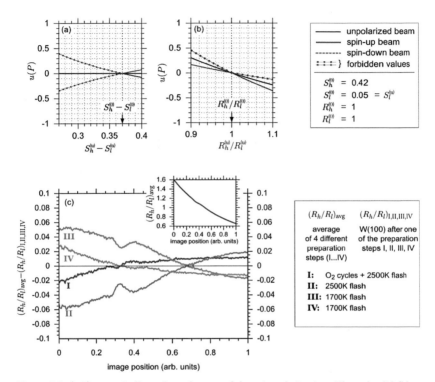

Figure 3.8: (a,b) numerically evaluated errors of the spin polarization. The red solid (blue dashed) line is the error for a fully polarized spin-up (spin-down) beam; the thinner black solid line represents an unpolarized beam. (a) errors arising through the deviation from the actual value of $S_h^{(0)} - S_l^{(0)}$. (b) errors arising through the deviation from the actual reflectivity ratio $R_h^{(0)}/R_l^{(0)}$ the values of R_l/R_h or S_h that correspond to zero error are indicated by arrows on the abscissa in every graph; "forbidden values" means that the value of the spin polarization $P^{(u)}$ obtained from analysis is above $+1.0$ or below -1.0. (c) Experimental variation of the reflectivity ratio R_h/R_l for four different conditions I, II, III, IV of the W(100) surface. The corresponding sequence of preparation steps is shown in the legend.

In figure 3.8(b), the influence of deviating R_h/R_l values is plotted, since the error of P depends only on the error of the ratio R_h/R_l (see equation (3.13)). The graph shows, that such an error leads to a shift of the obtained spin polarisation $P^{(u)}$ with respect to the true value $P^{(0)}$. The sign of the shift is independent of $P^{(0)}$. Too positive values of P mean, that spin-down intensity is erroneously ascribed to the spin-up channel. The opposite happens for too negative P.

In figure 3.8(c), line profiles of R_h/R_l across the field of view are plotted that have been obtained for several conditions of the W(100) surface resulting from different preparation steps. These have been measured as part of the experiments at the NanoESCA beamline of the synchrotrone ELETTRA, Trieste[3]. The R_h/R_l values vary considerably. Hence, spin-polarized measurements that have been performed after a particular preparation of the W(100) surface, have to be analyzed using R_h/R_l images obtained after the same preparation procedure. For properly determined R_h/R_l we estimate a remaining uncertainty of $u(R_h/R_l)=\pm 0.01$. Via figure 3.8(b) this translates to $u(P)\approx\pm 5\%$. Similarly, a remaining uncertainty of $u(S_h - S_l) = \pm 0.01$ translates to $u(P)\approx\pm 5\%$. Figure 3.9(a) shows that within 2 hours of measurement, the spin-related asymmetry varies by $\approx 10\%$ of the initial value. We estimate, that the error can be reduced to $u(S_h - S_l)\approx\pm 2\% \cdot 0.37\approx\pm 0.01$, if a suitable average value of $S_h - S_l$ is used. This value should be adapted according to the start time and end time of the spin-resolved measurement with respect to the preparation of the W(100) surface.

3.3.5   Time behaviour of the spin sensitivity of the W(100) crystal

The preparation of a clean W(001) surface has been described in [151]. We typically used 5 to 10 cycles of low-power flashes in oxygen atmosphere at a pressure of 5×10^{-8} mbar. The flashes (rapid heating for about $8\ldots 15$ seconds) were done by electron bombardment with an emission current of 55 mA and 1.2 kV voltage between filament and W(001) crystal. The maximum temperature during the flash was 1700 K as measured with an optical pyrometer. After the pressure returned to the lower 10^{-10} mbar range, a single high-temperature flash (2500 K) was applied using an emission current of 190 mA and a voltage of 1.15 kV.

Figure 3.9 shows the asymmetry of intensities recorded from two magnetic domains of an 8 monolayer Co thin film with a scattering energy of 26.5 eV at the W(001) crystal. In graph (a), the asymmetry drops to 65% of the initial value after a period of 9 hours. In graph (b), it is demonstrated that the spin sensitivity may be partially recovered by a low-temperature flash (without oxygen). This can be used to avoid the more time consuming preparation with oxygen flash cycles and subsequent high-power flash. With every low-power flash, however residual carbon contamination of the bulk tungsten crystal diffuses to the surface and will eventually have to be removed by the full preparation procedure. In our experiments, we have typically used 2 to 3 low-power flashes before repeating a full preparation.

The decrease of asymmetry shown in figure 3.9 is ascribed to the adsorption of hydrogen, which is known to change the work function of W(001) [152]. Since for low energies

---

[3]The electron lens settings used here correspond to larger ranges of polar and azimuthal angles of incidence. Consequently, the variation of the reflectivity ratio is also greater than shown in figure 3.6

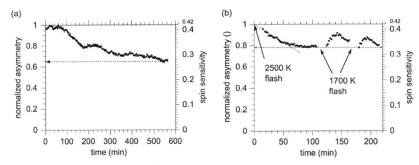

Figure 3.9: (a) change of the spin sensitivity in 10 hours after a high-power flash; (b) change of the spin sensitivity in 1-2 hours after a high-power flash and two low-power flashes. The asymmetry was calculated from the intensity values of two magnetic domains measured by two-photon photoemission in spatial mode and normalized to the value at $t = 0$.

between 20 eV and 100 eV, the universal curve for the inelastic mean free path predicts values below 0.6 nm [38], the scattering process is rather surface sensitive and could be affected by hydrogen adsorption. Numerous studies of hydrogen adsorption and desorption on W(001) have been performed [153–155] and it was found that most of the hydrogen is desorbed from the bcc(001) tungsten surface at temperatures between 400 and 700 K [155]. The fact, that a low-power flash can partially recover S, also points to the hydrogen as a cause for the change in S, since the temperature rises well above the desorption regime. Within $\approx$ 30 min after flashing, the temperature of the tungsten target falls back to room temperature and the adsorption sets in only after the temperature drops below $\approx$ 400 K. Additionally, the increased lattice constant at elevated temperature alters the electronic structure and consequently the spin-dependent scatter process results in a different value for the spin sensitivity. This explains, why the asymmetry in figure 3.9(b) approaches a maximum before it starts decreasing.

Figure 3.10 shows a time-dependent potential difference that was applied to the tungsten crystal with respect to the surrounding cylinder in order to compensate the continously changing work function of the W(001) surface in time. Without this correction, one observes an increasing lateral deflection and distortion of the scattered image. This is caused by an additional inhomogenous electric field occuring at the edges of the tungsten crystal which perturbs the electron beam path. The shown curve was empirically obtained, by minimizing the time-dependent image deflection and distortion that occurs during the photoemission experiments.

3.4   Co thin films on Cu(001)

The nearest neighbour distances of copper and cobalt are 2.556 Å(fcc lattice constant 3.615 Å) and 2.507 Å(corresponding to an "fcc lattice" constant of 3.544), respectively [156]. Thin Co films grow pseudomorphically on Cu(001), adopting its fcc structure.

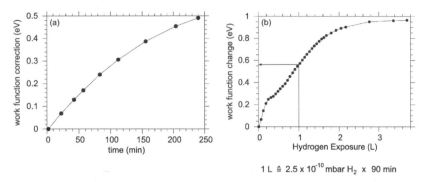

Figure 3.10: (a) change of the work function of the W(100) surface after a high- or low-power flash. The curve was empirically determined to minimize a time-dependent lateral deviation of the detected image. The pressure in the environment of the W(001) crystal was $\approx 1 \times 10^{-10}$ mbar. (b) change of the work function as a function of hydrogen exposure; data from Herlt [152].

Surface-extended X-ray absorption fine structure (EXAFS) measurements have shown a tetragonal distortion with a contraction of 4% (surface normal to inplane directions) for a thickness range from 2 to 15 monolayers, resulting in inplane and perpendicular lattice constants of $a = 3.61$ Å and $c = 3.47$ Å [157]. With these values, the fct Brillouin zone can be constructed in proper units to enable a comparison with the wave-vector radius of free-electron-like spheres when interpreting momentum-resolved photoemission measurements. Therefore, the base vectors of the reciprocal lattice are needed

$$\vec{b}_1 = \frac{2\pi}{a} \begin{pmatrix} 1 \\ 1 \\ -a/c \end{pmatrix} \quad \vec{b}_2 = \frac{2\pi}{a} \begin{pmatrix} 1 \\ -1 \\ a/c \end{pmatrix} \quad \vec{b}_3 = \frac{2\pi}{a} \begin{pmatrix} -1 \\ 1 \\ a/c \end{pmatrix}, \tag{3.26}$$

allowing a construction of the Wigner-Seitz cell in reciprocal space, viz. the Brillouin zone. The $\overline{\Gamma X}$ distances of the fct Brillouin zone for the Co layer are

$$\overline{\Gamma X}_\perp = 2\pi/c \approx 1.81 \text{ Å}^{-1} \quad \overline{\Gamma X}_{\parallel} = 2\pi/a \approx 1.74 \text{ Å}^{-1} \tag{3.27}$$

for surface-normal and in-plane directions, respectively.

In this work, Co films with thicknesses up to 18 ML were grown on Cu(001), where the growth rate calibration was performed by measuring the medium energy electron diffraction (MEED) intensity as a function of time. Auger electron spectroscopy has been used to test the homogeneity of the thickness across the surface area of the Cu(001) single crystal.

From measurements of the surface Kerr magneto-optical effect, the Curie temperature is known to vary as a function of thickness from 100 K (at 1.5 ML) to 600 K (at 3 ML) [158], while the value for bulk hcp Co is 1117°C (1390K) [159]. Furthermore, the remanent magnetization was found in-plane for 1.5 to 20 ML with the easy axes pointing along the <110> directions (for $d_{Co} > 2$ ML). In our experiments, we chose a growth temperature of 410 K which is above the Curie temperature up to a thickness of 2.2 ML according to [158].

Chapter 4

Measurements

4.1 Fermi surface of fct cobalt mapped by constant-initial state photoemission

Photoemission measurements were carried out at the NanoESCA beamline of the synchrotron ELETTRA in Trieste[1]. It provides a commercial photoemission microscope (Omicron NanoESCA [135]). A schematic was shown in figure 3.1(a) (red outlined parts are not part of the permanent installation).

Before we present measurements of the valence band structure of Co thin films, it is helpful to introduce the Fermi surface sheets predicted by theory. Figures 4.1(a-d) show the Fermi surface of fcc Co from calculations using tight-binding models [160, 161] (similar results published earlier by another group in [15]). The parts shown in (a-d) are resolved by majority and minority spin and further by symmetry properties of the wave functions. We introduce the symbols $S_\mathcal{A}^\downarrow$, $S_\mathcal{B}^\downarrow$, $S_\mathcal{C}^\downarrow$, $S_\mathcal{D}^\uparrow$ for further reference. Figure 4.1(e) is a planar cut through the shapes plotted in (a-d), and shows the Fermi level crossings along the Δ axis of the fcc Brillouin zone, which are marked by arrows. A fully relativistic band structure calculation of fcc cobalt [8] is shown in figure 4.1(f). The Fermi level crossings here can be related to figure (e). Thereby, the shapes in (a-d) can be assigned to the individual bands in (f), as well. Accordingly, $S_\mathcal{D}^\uparrow$ is linked to the Δ_1 majority band, while $S_\mathcal{A}^\downarrow$ is linked to the $\Delta_1^\downarrow$ band. This is consistent with the sp-assignment of the $S_\mathcal{A}^\downarrow$ in [15],

Both $S_\mathcal{B}^\downarrow$ and $S_\mathcal{C}^\downarrow$ can be associated with the $\Delta_5^\downarrow$ band crossing, since the $\Delta_5^\downarrow$ band is two-fold degenerate, i.e. these two minority Fermi surface sheets touch at the point where they intersect the Δ axis. This is apparent in figure 4.1(e). The Δ_5 degeneracy is lifted when deviating from the Δ axis, so that two Fermi surface sheets are formed. The part of $S_\mathcal{C}^\downarrow$ facing the Γ point in fig. 4.1(c) follows to be predominantly of Δ_1 character, since the $\Delta_1^\downarrow$ and the $\Delta_5^\downarrow$ bands are hybridized as shown in the band structure and the $\Delta_1^\downarrow$ band intersects the Fermi level closer to the Γ point than all other bands.

4.1.1 Photon energy scan

In order to probe the Fermi surface throughout the Brillouin zone, we performed a constant-initial-state measurement with $E_f = h\nu - \phi \Rightarrow E_i = E_F$. The photon en-

[1]endstation hosted by the electronic properties group (director C.M. Schneider) of the Peter-Grünberg Institute at Forschungszentrum Jülich

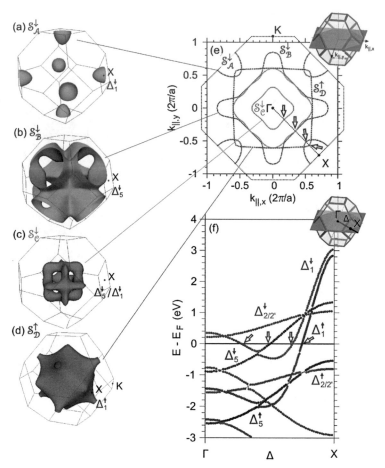

Figure 4.1: (a-d) Theoretically computed spin- and orbital resolved Fermi surface of fcc Co (Data from [160, 161]). Majority and minority spins are indicated by red upwards arrows and blue downward arrows, respectively. (a) has been assigned to an sp-band and encloses unoccupied states while (b-c) have been assigned to d-bands in [15]. (d) is of majority spin type (e) planar cut through the orbital and spin-resolved Fermi surface sheets in (a-d), the $k_{||,x}, k_{||,y}$ axes correspond to momentum-resolved photoemission images from a (100) surface. (f) fully-relativistic band structure calculation of fcc cobalt along the Δ axis (data from [8])

ergy has been increased in 1-eV-steps covering a range from 35 eV to 200 eV. At each photon energy, momentum-resolved images were recorded for a 1 eV range of initial state energies centered around E_F. These momentum-integrated binding energy spectra are dominantly determined by Fermi statistics, i.e. they exhibit a step, which was modelled by an error function. The fitting was used to extract the position of the Fermi level for every photon energy. For the resulting value of $E_f = E_F + h\nu$, the $(k_{||,x}, k_{||,y})$ image was linearly interpolated from two images at $E_{1,2}$ ($E_1 < E_f < E_2$, $E_2 - E_1 \approx 50$ meV).

$$I_{E_F}(k_{||,x}, k_{||,y}) = I_1(k_{||,x}, k_{||,y}) + \frac{E_1 - E_F}{E_1 - E_2}(I_2(k_{||,x}, k_{||,y}) - I_1(k_{||,x}, k_{||,y})) \qquad (4.1)$$

Repeating this procedure for every photon energy, we obtain the Fermi level initial-state photoemission intensity I_{E_F} as a function of three variables $k_{||,x}, k_{||,y}, h\nu$. Figure 4.2 shows two subsets of this data including the $h\nu$ axis and one of two selected $k_{||}$ axes which are indicated by the planar intersection of the Brillouin zones shown to the right of the graphs.

The theoretical Fermi surface in figures 4.1(a-d) was mapped as follows: Each point within the first Brillouin zone $\vec{k}_0 = (k_{||,x,0}, k_{||,y,0}, k_{\perp,0})$ that belongs to the Fermi surface, can be rewritten $\vec{k}_0 = \vec{k}_j + \vec{G}_j$ where $\vec{G}_j$ are reciprocal lattice vectors of the fcc lattice. $\vec{k}_j + \vec{G}_j$ is inserted in equation (2.6). After replacing E_f by $E_F + h\nu_j$, one can solve for $h\nu_j$. The resulting points $(k_{||,x,j}, k_{||,y,j}, h\nu_j)$ were plotted in figure (4.2). During this mapping, the inner potential U_i was a free parameter and best agreement with our experimental data was achieved by choosing $U_i = 12.3$ eV $+ \phi$.

Two measurements with freshly prepared Co thin films have been made, each with a duration of 8 hours and separating the total photon energy range at $h\nu = 70$ eV. The pressure was 1×10^{-10} mbar. The order of the photon energies during each measurement was ascending. Exposure times were in the range 15 s to 180 s (for $h\nu > 71$ eV, $t \leq 30$ s).

Best agreement between theory and experiment is found for $\mathcal{S}_{\mathcal{D}}^{\uparrow}$ in figure 4.2(a). Two photoemission resonances running vertically through the graph at $|k_{||}| = \pm 1.2 \text{Å}^{-1}$ form a bottle-like shape. They overlap with the vertical sections of $\mathcal{S}_{\mathcal{D}}^{\uparrow}$. From theory, the necks at the L point in figure (a) appear larger than in the experimental data, where the high-intensity feature due to majority Fermi surface sheets from neighbouring Brillouin zones forms a rather connected line.

More similarities between experiment and theory are found for $\mathcal{S}_{\mathcal{C}}^{\downarrow}$ in the photon energy range (35 eV $< h\nu <$ 60 eV). The shape of the high-intensity feature is less extended along the $h\nu$ axis than predicted by theory. We also find high-intensity features corresponding to the hole pockets $\mathcal{S}_{\mathcal{A}}^{\downarrow}$ in figure 4.2(b) (35 eV $< h\nu <$ 60 eV), while at $h\nu \approx 100$ eV, they are either absent or very weak. For $\mathcal{S}_{\mathcal{B}}^{\downarrow}$, no enhanced intensity is found in figure (a), but in figure (b) the shape of $\mathcal{S}_{\mathcal{B}}^{\downarrow}$ is approximately reproduced by the measurement (60 eV $< h\nu <$ 80 eV), although with a shift in $h\nu$. On the other hand, the majority Fermi surface sheet $\mathcal{S}_{\mathcal{D}}^{\uparrow}$ in figure (b) shows a shift in the opposite direction. This could be due to a smaller size of the volume enclosed by $\mathcal{S}_{\mathcal{B}}^{\downarrow}$ than predicted by theory. For the photon energies 85 eV and 125 eV, the enhanced intensity around $k_{||} = 0$ may be ascribed to $\mathcal{S}_{\mathcal{A}}^{\downarrow}$ or $\mathcal{S}_{\mathcal{D}}^{\uparrow}$, but they are further apart on the energy axis than the extension of both Fermi surface sheets along $k_{||} = 0$. Either these partial surfaces enclose a bigger volume, or — since we are in the vicinity of the Brillouin zone boundary — the final state band deviates from the parabolic dispersion and opens up a gap, where no direct transitions are possible.

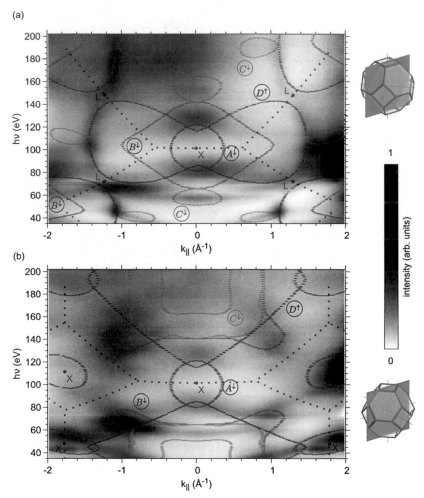

Figure 4.2: Constant-initial state measurement of the Cobalt Fermi surface for a photon energy range from 35 eV to 200 eV. The intensity has been normalised to the mean value for every photon energy, separately. Along the $k_{||}$-axis, intensity values are consistent. A non-linear gray scale was used for emphasis of the weaker resonances. The black dotted lines are the fct Brillouin zone boundaries. The colored points are a mapping of the theoretical Fermi surface sheets [160, 161] to a $h\nu$-scan using equation (2.5) with an inner potential $U_i = 12.3$ eV $+\phi$ and an effective mass $m_{\mathrm{eff}}/m_{\mathrm{e},0} = 1$. Red (blue) points belong to the majority (minority) parts of the Fermi surface. High symmetry points of the Brillouin zone are marked with filled black circles. (a) $k_{||}$-axis along the (110) direction (containing the in-plane projection of the L point) (b) $k_{||}$-axis along the (100) direction (containing the in-plane projection of the X point)

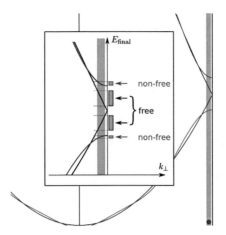

Figure 4.3: direct transition from an initial state within some $k_\perp$-range (red circle) close to the Brillouin zone boundary to (1) a free-electron final state band (black solid line, equation (2.5)) or (2) a nearly-free electron final state band (blue solid line) with a gap at the Brillouin zone border. The blue and black arrows point to ranges of photoelectron energy where the highlighted $k_\perp$ range is mapped to by direct transitions.

Such a situation is illustrated in figure 4.3. Assuming a non-free (or nearly-free) type of electron dispersion for the final state bands, electronic states within a $k_\perp$-range close to a Brillouin zone boundary are mapped to intervals of photoelectron energies that are farther away from $E(\vec{k} = X)$ than if a free-electron dispersion was used. This is simply a consequence of introducing a band gap into the final state bands that pushes energy eigenvalues away from the energy value where the free electron parabola touches the Brillouin zone border. A second effect is the compression of the energetic interval that the $k_\perp$-range is mapped to, compared to the free-electron case (blue-contour boxes versus black-contour boxes). Yet another consequence is the formation of an intensity gap in the photon energy axes in figures 4.2(a,b) around the X point ($h\nu$=100 eV, $k_{||}$=0), where due to the lack of final states, resonant transitions are absent.

Overall, the best agreement is found for the majority Fermi surface sheet $S_D^\uparrow$. The corresponding photoemission intensity is strongly enhanced and discernable throughout the complete photon energy range $35\,\text{eV} < h\nu < 200\,\text{eV}$. A deviation from theory is found for the neck at the L point which has a smaller k-diameter in experiment. For every minority Fermi surface sheet $S_A^\downarrow$, $S_B^\downarrow$, $S_C^\downarrow$, correspondance to high-intensity features was found for some photon energy ranges, while for some other energy ranges, where the $k_\perp \rightarrow h\nu$ mapping predicts their presence, they are too weak for observation (or absent).

4.1.2 Momentum distributions at discrete photon energies

In figure 4.4(a), we show a $(k_{||,x}, k_{||,y})$ image of photoemission probing the Fermi surface with a photon energy of 42 eV. The azimuth of the incidence of light on the sample surface

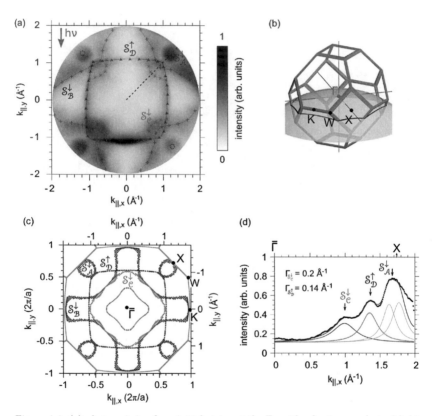

Figure 4.4: (a) photoemission from initial states at the Fermi level using p-polarized light with $h\nu = 42$ eV. Photoemission peak positions are marked by crosses/circles. (b) Bulk Brillouin zone (blue) for a face-centered cubic structure. The yellow surface represents the spherical constant-energy surface according to equation (2.5) with $E_{\text{kin}} + U_i = 54$ eV. (c) parallel-momentum projection of the spherical cut (b) through the theoretical surface (figure 4.1(a-d)). The grey outline indicates the border of the fct Brillouin zone. (d) photoemission-intensity line profile between $\bar{\Gamma}$-point (normal photoemission) and X point.

is indicated by an arrow in the top left corner. The shapes of high-intensity features respect the symmetry of the (001) surface, which consists of four-fold rotational symmetry as well as 2 inequivalent mirror operations (with mirror planes that contain the (100) direction or the (110) direction). Figure 4.4(b) is the constant final state energy surface defined by equation (2.6), with $U_i = 12.3\,\text{eV} + \phi$ (obtained in section 4.1.1) and $\vec{G} = (0, 0, -4\pi/c)$; c as defined in equation (3.27). It contains the high symmetry axis $\Gamma - X$ apart from a certain deviation due to the curvature: The border of the Brillouin zone in the vicinity of the high-symmetry points X, W and K is intersected at $k_\perp = -0.16\,\overline{\Gamma X}$, while the $k_\perp$ axis is intersected $0.08\,\overline{\Gamma X}$ units above the Γ point.

The points in figure 4.4(c) are obtained from the intersection of the surface shown in figure 4.4(b) with the theoretical Fermi surface in figure 4.1(a-d). Their comparison with the shapes in figure 4.4(a) leads to the assignment of the photoemission resonances to the respective $\mathcal{S}^{\uparrow\downarrow}$. We find good agreement in case of $\mathcal{S}_A^\downarrow$ and $\mathcal{S}_D^\uparrow$ and less exact agreement for $\mathcal{S}_B^\downarrow$ and $\mathcal{S}_C^\downarrow$. A line profile between the center (within the surface Brillouin zone denoted by $\overline{\Gamma}$ which is equivalent to normal emission) and the X point as shown by the dashed straight line in figure 4.4(a) is shown in figure 4.4(d). Three distinct intensity peaks have been modelled by Lorentz peaks with a subsequent Gaussian convolution (FWHM 0.047 Å^{-1}) to include the effect of instrumental broadening. The line profile shows that the intensity feature in figure 4.4(a) marked by $\mathcal{S}_C^\downarrow$ is a separate resonant structure. However, the inner contour of $\mathcal{S}_C^\downarrow$ (see the innermost contour enclosing the Γ point in figure 4.4(c)) does not correspond to any peak in the line profile.

The Lorentzian Γ parameters (equivalent to half width at half maximum) are given in the graph (d) for the peaks connected to $\mathcal{S}_C^\downarrow$ and $\mathcal{S}_D^\uparrow$. For the peak corresponding to $\mathcal{S}_A^\downarrow$, the line shape consists of two Lorentz peaks as expected from figure 4.4(c).

Figure 4.5 shows momentum-resolved photoemission images for selected photon energies. The area of probed wave vectors is illustrated by the Brillouin zone sketches below each graph. The intersection of the final state surface with the boundary of the Brillouin zone is marked by red lines drawn in the momentum-resolved images. Although the Brillouin zone border is symmetric with respect to a mirror operation $k_z \to -k_z$, the red lines show a convex deformation for $h\nu \leq 42\,\text{eV}$ and $h\nu > 100\,\text{eV}$ while for $42\,\text{eV} < h\nu < 100\,\text{eV}$, the deformation is concave. These parts of the Brillouin zone border belong to the hexagonal faces which contain the L points and the deformation arises due to the intersection with a spherical surface. The photoemission patterns follow this concave/convex characteristics, as can be seen for $h\nu = 80\,\text{eV}$ and $h\nu = 130\,\text{eV}$.

In these momentum images, more than one Brillouin zone is probed, i.e. the photoelectron receives a crystal momentum (for normal emission and the adjacent $k_{||}$-area):

$$\hbar\vec{G}_{0,n} = 2\pi\hbar \left(0, 0, \frac{2n}{c}\right)^\top \tag{4.2}$$

or (in case of the regions outside the centered Brillouin zone which are marked by the red lines)

$$\hbar\vec{G}_{1,n} = 2\pi\hbar \left(\frac{\pm 1}{a}, \frac{\pm 1}{a}, \frac{1 + 2n}{c}\right)^\top \tag{4.3}$$

The adjacent regions probe Brillouin zones which are shifted in $k_\perp$ direction by a $\overline{\Gamma X}$

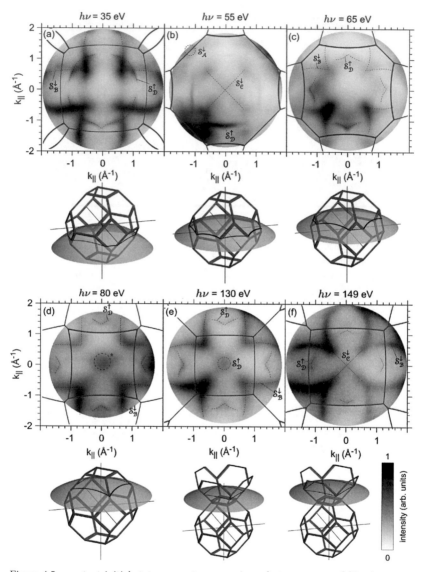

Figure 4.5: constant initial state momentum mapping: photon energy and kinetic energy of the photoelectrons are varied by the same amount to probe initial states at the Fermi level at different $k_\perp$ values. The figures below each measurement show the fct Brillouin zone with a plane of constant final state energy where the final state is approximated as a free electron. Its energy has been calculated as $E_f = h\nu - \phi + U_i$ with the inner potential set to $U_i = 12.3$ eV $+ \phi$.

distance. Hence, the high-intensity features in the momentum images shown here are generally not mirror symmetric with respect to the Brillouin zone border. Exceptions are the X points in the $k_\perp = 0$-plane which can be probed with $h\nu = 50\,\text{eV}$ and the L points which can be probed with $h\nu = 70\,\text{eV}$.

Most of the observed high-intensity features in figure 4.5 can be traced back to the theoretical Fermi surface and hence, to the bulk band structure of fcc Co. For example, the nearly quadratic cross section of $\mathcal{S}_{\mathcal{D}}^\uparrow$ seen in figure 4.4(a) and 4.5(b) turns into a $45°$ rotated object with corners along the $k_{\parallel,x}$ and $k_{\parallel,y}$ axes in figure 4.5(c,f). In figure (d,e), these corners can only be seen outside the central Brillouin zone. The top of $\mathcal{S}_{\mathcal{D}}^\uparrow$ causes a high-intensity feature at normal photoemission in figure 4.5(d).

A lot of the well discernible photoemission resonances are linked to $\mathcal{S}_{\mathcal{B}}^\downarrow$. The intersection of $\mathcal{S}_{\mathcal{B}}^\downarrow$ with final-state spheres either resembles the closed shape shown in figures 4.1(e), 4.4(a), 4.5(a); or it forms separated "corners" as found in figures 4.5(c,d,e). Its associated electronic band has a rather large anisotropy, since the high-intensity feature linked to $\mathcal{S}_{\mathcal{B}}^\downarrow$ (figure 4.1(d)) has a peak in the bulk (100) directions (diagonals of the momentum image) and no peak in the (110) directions ($\mathcal{S}_{\mathcal{B}}^\downarrow$ forms a wide neck at the L points).

$\mathcal{S}_{\mathcal{C}}^\downarrow$ corresponds to a hybridized band in figure 4.1(e) that crosses the Fermi level twice. Hence it has inner and outer faces pointing to the Γ point or to the border of the Brillouin zone. In figure 4.4, we found at most a part of the outer contour, while the inner is either missing or not discernible from the photoemission background. For some higher photon energy, we expect to observe a cross-shaped feature, as seen for the part of $\mathcal{S}_{\mathcal{C}}^\downarrow$ that faces the front X point in figure 4.1(c). The bars of the cross are oriented towards in-plane projections of the X points, which corresponds to the diagonal directions in our momentum images. We observe such patterns in figures 4.5(b,f).

$\mathcal{S}_{\mathcal{A}}^\downarrow$ is found in form of circular features located at the Brillouin zone border for $h\nu=42\,\text{eV}$ and less strongly for $h\nu-55\,\text{eV}$. From the band structure shown in figure 4.4(e), the k-radius of these features grows towards energies below the Fermi level.

In summary, good agreement between the calculated Fermi surface from literature and our constant-initial state photoemission data was found and has lead to a value for the inner potential $E_{\text{F}}-U_{\text{i}}=12.3\,\text{eV}$ which is used for equation (2.6). Furthermore, the measurements shown here provide a link between binding energy spectra obtained at different photon energies. In view of the electronic correlation, we add that the calculated Fermi surface which we used for comparison to our experimental data, corresponds to a single-particle picture and neglects the electron self energy Σ. For the imaginary part, this is acceptable, since the lifetime broadening due to electron-electron interaction always vanishes at the Fermi level. The real part of the electron self energy may be important and lead to a shift of quasiparticle bands, including the ones which cross the Fermi level. Depending on the slope of such bands at $E=E_{\text{F}}$, this shift results in an increase or decrease of the volume of their associated part of the Fermi surface With respect to Luttingers theorem[162], the overall volume enclosed by the Fermi surface is not changed by interactions with respect to a non-interacting system. However, the individual parts of the Fermi surface connected to different bands that cross the Fermi level may "exchange" k-space volume among each other.

4.2 Valence electronic structure of Co/Cu(001)

4.2.1 Valence electronic structure in the ΓWXK plane

Photoemission results using p-polarized radiation ($h\nu$=50 eV) are presented in figure 4.6. The high-symmetry points X, K and W are quite accurately contained within the surface of probed wave vectors (see figure 4.7(a,b)). Their positions are indicated in the momentum-resolved image in figure 4.6(a) and in the energy spectra in figures 4.6(d,e,f). Accordingly, the dispersion of the electronic states along the X-W axis is obtained by a simple E_B-scan and can be compared to calculated data in [163] - see figure 4.6(e).

For the high-intensity feature P_4, we find good agreement with regards to the slope of the dispersion, although the calculated majority and minority band deviate in form of a rigid energy offset. In figure 4.6(a), the corresponding feature appears as a circular shape centered around each of the four X points. From a comparison with figure 4.4(a), the structure in figure 4.6(a) along X-W belongs to the part $S_{\mathcal{A}}^{\downarrow}$ of the minority-spin Fermi surface. Its wave function for a Bloch wave vector on the Δ axis belongs to the Δ_1 symmetry group, which transforms like the constant scalar function "1". We find an increasing k-radius of these circular shapes towards lower energies (compare figures 4.6(a) and (b)), which is consistent with the slope of the Δ_1 minority band in the theory data of figure 4.4(f)).

A further high-intensity feature P_5 is observed at E_i=E_F−0.2 eV in figure 4.4(e) and at the X points in figure 4.4(f). It extends along the X-W edge of the Brillouin zone border and is rather broad (0.4 Å^{-1}) in the $k_\parallel$ direction perpendicular to its extension (i.e. along the $X-\bar{\Gamma}-X$ shown in figure 4.4(f)). No equivalent valence band is found in the calculated electronic structure of [163].

The dispersion of the rectangular resonance P_1 in figure 4.6(c) which is associated with $S_{\mathcal{D}}^{\uparrow}$, can be seen as a strongly resonant feature down to an energy $E-E_F$=−0.3 eV. The minimum is found for −0.7 eV. In the range between −0.3 eV and −0.7 eV, the amplitude of the resonance decreases strongly and the FWHM of the peak in a constant-energy $k_\parallel$-lineprofile increases distinctly.

Another high-intensity feature P_2 extends across the energy range E_F−1.5 eV$<E_i<E_F$ in figure 4.6(d) ([110] direction), partially outside the central Brillouin zone. Its minima are at $k_\parallel$=±1.1Å^{-1} and E_i=E_F−1.5 eV. From figure 4.2(a) follows, that it is associated with $S_{\mathcal{B}}^{\downarrow}$. In the momentum-resolved image for E_i−E_F=−1 eV, this high-intensity feature appears four times on the horizontal and the vertical axis, respectively. Two of these peaks have been marked by arrows to relate them to each other in figures 4.6(b) and (d).

In the theory data [163], we find a band that is slightly shifted in energy but shows a minimum at the same $k_\parallel$ position as the photoemission resonance. The photoemission resonance P_3 may be due to an exchange-split opposite-spin band of the $S_{\mathcal{B}}^{\downarrow}$-related structure P_2, since both have an energy minimum at $k_\parallel$=±1.1Å^{-1}. The band width of P_3 is found to be smaller than for P_2. If these are spin-split bands, the band narrowing of P_3 may be due to a stronger electron-electron interaction for the lower band [164]. The spin polarization of these photoemission resonances is presented in section 4.2.2.

Figure 4.6(f) shows photoemission data along the (in-plane) [100] direction. A multitude of peaks are observed in the rather narrow energy range E_F−1 eV$<E_i<E_F$, whereas no

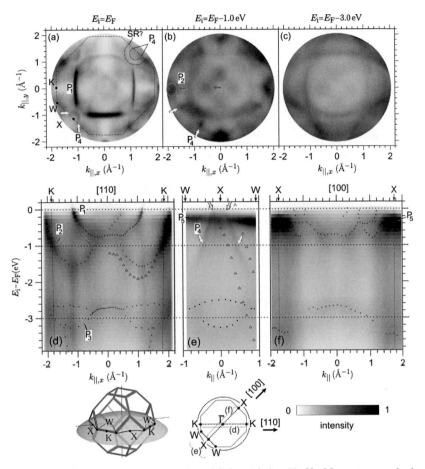

Figure 4.6: photoemission using p-polarized light with $h\nu$=50 eV. Momentum-resolved photoemission images probing initial states at (a) E_F−3 eV, (b) E_F−1 eV, (c) E_F. The dotted outline in (c) indicates the Brillouin zone border. (d-f) $E(k_{||})$ images with $\vec{k}_{||}$ as indicated below figures (d,e) and by the labelled high symmetry points X,W,K on top of the graphs. Triangles in (e) are theoretical data from [163]. Blue and red indicate minority and majority spin. Gray dots in (d,f) are guides to the eye for local intensity maxima (positions obtained by the criterium of vanishing gradient and negative laplacian of the intensity). Arrows indicate corresponding peaks in energy and momentum-resolved images.

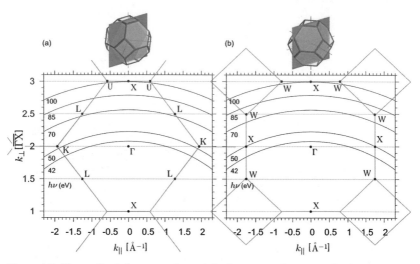

Figure 4.7: Planar $(k_{||}, k_\perp)$ cross sections of the face-centered tetragonally distorted bulk Brillouin zone for each of the inequivalent mirror planes of the (001) surface. (a) $k_{||}$ along the (110) direction (b) $k_{||}$ along the (100) direction The green outline indicates the borders of several Brillouin zones. The black curved lines show the constant final state surface for various photon energies and zero binding energy (initial states at Fermi level) according to equation (2.6) with $U_i = 12.3$ eV $+\phi$, $\vec{G} = (0, 0, 4\pi/c)$.

features due to dispersing bands are observed in the energy range $E_F - 2.5$ eV $< E_i < E_F - 1$ eV. Instead, we find vertical bars of intensity maxima. Line profiles in figure 4.8 show in detail the part of the spectrum, where the peak positions are found at constant $k_{||} = \pm 0.35$Å^{-1}, independent of the binding energy. We rarely find any further resonant features that correspond to dispersive quasiparticle bands. This characteristic of photoemission spectra is rather uncommon and will be discussed in section 5.1.

4.2.2   Spin-resolved results

In preparation of the spin-resolved measurements, the sample has been homogeneously magnetized parallel to the spin-sensitive axis in the direction 'M+', which is defined by the polarity of an electric current pulse driven through a magnetizing coil. The homogeneity of the magnetization in the cobalt film was verified in the spin-resolved spatial imaging mode within an area of $\approx (80\,\mu\text{m})^2$. This is sufficiently larger than the area probed during the momentum-resolved measurements, which is determined by the beam diameter ($\approx 20\,\mu\text{m}$). Afterwards, the procedure detailed in section 3.3.2 has been performed to obtain spin-resolved $(k_{||,x}, k_{||,y})$ images for selected binding energies. The results are presented in figure 4.9 as plots of the majority-spin and minority-spin partial intensities.

For initial states close to the Fermi level, we find a strong separation of photoemission maxima into majority-spin channel and minority-spin channel (figures 4.9(a,b)), i.e. nearly

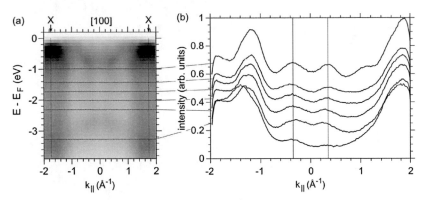

Figure 4.8: $k_{||}$-lineprofiles along the (100) direction for the energy region $E_F-2.5\,\mathrm{eV}<E_i<E_F-1.0\,\mathrm{eV}$, where the spectrum in figure 4.6(f) shows vertical bars of high photoemission intensity.

fully polarized peaks. The rectangular shape ($\mathcal{S}_D^\uparrow$) in the majority-spin ($k_{||,x},k_{||,y}$) image is accompanied only by a low intensity enhancement in the minority-spin image. The same is found for the resonances along the W-X line in figure 4.9(b,c). Contrarily, the photoemission features at $E_i=E_F$, previously assigned to $\mathcal{S}_A^\downarrow,\mathcal{S}_B^\downarrow,\mathcal{S}_C^\downarrow$, show a dominating minority-spin polarization.

In figure 4.9(e), we observe a high-intensity feature most distinguishable in the minority channel, which is connected to the U-shaped bands in figure 4.9(d) (minima at $k_{||}\approx\pm1.2\,\text{Å}^{-1}$ and $E_F-1.7\,\mathrm{eV}$). The spin polarization measured on this resonance is actually positive, which, however, seems to be caused by a higher average majority spin intensity of the entire ($k_{||,x},k_{||,y}$) field of view, not specific to this peak. This motivates a method to distinguish the spin polarization of a photoemission peak (Peak) from the spin polarization of the background (BG). The partial intensities can be written as a sum, with corresponding spin polarization values as follows:

$$I_{\uparrow,\downarrow} = I_{\uparrow,\downarrow,\mathrm{Peak}} + I_{\uparrow,\downarrow,\mathrm{BG}} \tag{4.4}$$

$$P = \frac{I_{\uparrow,\mathrm{Peak}} - I_{\downarrow,\mathrm{Peak}} + I_{\uparrow,\mathrm{BG}} - I_{\downarrow,\mathrm{BG}}}{I_\uparrow + I_\downarrow} \tag{4.5}$$

$$P_{\mathrm{Peak}} = \frac{I_{\uparrow,\mathrm{Peak}} - I_{\downarrow,\mathrm{Peak}}}{I_{\uparrow,\mathrm{Peak}} + I_{\downarrow,\mathrm{Peak}}} \qquad P_{\mathrm{BG}} = \frac{I_{\uparrow,\mathrm{BG}} - I_{\downarrow,\mathrm{BG}}}{I_{\uparrow,\mathrm{BG}} + I_{\downarrow,\mathrm{BG}}} \tag{4.6}$$

$$P = P_{\mathrm{BG}}\frac{I_{\uparrow,\mathrm{BG}} + I_{\downarrow,\mathrm{BG}}}{I_\uparrow + I_\downarrow} + P_{\mathrm{Peak}}\frac{I_{\uparrow,\mathrm{Peak}} + I_{\downarrow,\mathrm{Peak}}}{I_\uparrow + I_\downarrow} \tag{4.7}$$

The background intensity $I_{\uparrow,\downarrow,\mathrm{BG}}$ is caused (1) by inelastically scattered electrons, which contribute as a homogeneous intensity in the ($k_{||,x},k_{||,y}$) field of view, and (2) by an ensemble of neighbouring photoemission peaks (which may be broader than the resonance of interest). The resulting P_{Peak} may have a different sign than the measured P. The remaining issue is, how to extract the spin-resolved peak intensity $I_{\uparrow,\mathrm{Peak}}, I_{\downarrow,\mathrm{Peak}}$ with

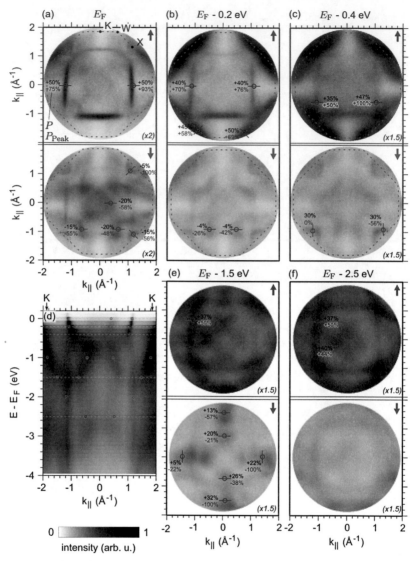

Figure 4.9: (a-c,e,f) spin-resolved photoemission using p-polarized light with $h\nu = 50\,\text{eV}$. Up (down) arrows indicate a majority spin ($I_\uparrow$) or minority spin ($I_\downarrow$) partial intensity spectrum; (d) data from figure 4.6(d) for reference, horizontal lines are drawn at the energies corresponding to figures (a-c,f); numbers in parentheses give the factor, by which the gray scale intensity was multiplied; a consistent intensity gray scale has been used for every pair of $I_\uparrow$ and $I_\downarrow$ spectrum. The values P and P_{Peak} (see text) are given for selected points, which are indicated by red and blue circles The line profile for obtaining P_{Peak} was plotted along the direction indicated by the bars;

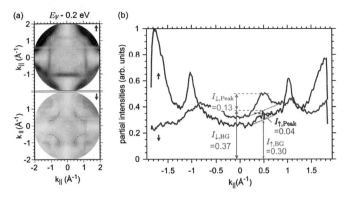

Figure 4.10: $k_{||}$-line profile of the spin-resolved spectrum shown in figure 4.9(b)

which P_{Peak} can be calculated according to equation (4.6). For the data in figure 4.9, values of $I_{\uparrow,\text{BG}}$, $I_{\downarrow,\text{BG}}$ have been obtained as illustrated in figure 4.10. The error of $u(P_{\text{Peak}})$ can be estimated as

$$\frac{u(P_{\text{Peak}})}{u(P)} = \frac{I_{\uparrow,\text{Peak}}+I_{\downarrow,\text{Peak}}+I_{\uparrow,\text{BG}}+I_{\downarrow,\text{BG}}}{I_{\uparrow,\text{Peak}}+I_{\downarrow,\text{Peak}}} \tag{4.8}$$

For the example in figure 4.10 , we obtain $P=-0.2\pm0.10$ and $P_{\text{Peak}}=-0.5\pm0.32$. $u(P_{\text{Peak}})$ increases as the ratio of peak intensity to background intensity decreases.

Values of P and P_{Peak} are printed for various majority-spin and minority-spin resonances in figure 4.9. Using the sign of P_{Peak} as a criterium, we have indicated dominating majority spin (red circles) or minority spin (blue circles) character to the photoemission maxima in figure 4.9(d). Continuing the discussion of the U-shaped bands in figure 4.9(d) ($E_{\text{F}} - <E_i<E_{\text{F}}-1.7\,\text{eV}$), we note, that they are a dominantly minority-spin feature on a majority-spin background. A similar, although more broadened feature in the range $E_{\text{F}}-3.5\,\text{eV}<E_i<E_{\text{F}}-2.5\,\text{eV}$) is likely the majority-spin complement, which would correspond to an exchange splitting of $\approx(1.8\pm0.2)\,\text{eV}$.

The non-dispersive photoemission resonance at constant $k_{||}-\pm1.1\,\text{Å}^{-1}$ is found to have the same value of P as the photoelectrons of nearby ($k_{||,x},k_{||,y}$) points (observable in an image $P(k_{||,x},k_{||,y})$ which is not shown here). According to equation (4.7), $P_{\text{BG}}=P$ implies $P_{\text{BG}}=P_{\text{Peak}}$, i.e. this resonance does not contribute with a specific spin polarization.

The $|k_{||}|<2\,\text{Å}^{-1}$-average of the spin polarization is dominantly of majority type for $E_{\text{F}}-1\,\text{eV}<E_i<E_{\text{F}}-4\,\text{eV}$. This majority spin intensity is not concentrated in $k_{||}$-localized peaks, but evenly distributed over the field of view. The fact, that close to the Fermi level, we find a high density of minority spin electronic states speaks against an interpretation in terms of inelastically scattered photoelectrons. Secondary electrons with kinetic energies larger than 5 eV from hcp Co were reported to have a spin polarization of $+19\%$ independent of the exciting photon energy ($20\,\text{eV}<h\nu<110\,\text{eV}$) [165]. We find $k_{||}$-integrated spin polarization values larger than $+19\%$ as listed in table 4.1. The conclusion is, that the observed broadening in the ($k_{||,x},k_{||,y}$)-images of the majority spin photocurrent

$E - E_{\mathrm{F}}$	-0	-0.1	-0.2	-0.3	-0.4	-0.5	-0.8	-1.3	-1.8	-2.3	-2.8
$P(\%)$	0	+14	+28	+34	+34	+33	+32	+20	+26	+23	+17

Table 4.1: spin polarization $P^\diamond$ calculated as $P^\diamond=(I_\uparrow^\diamond - I_\downarrow^\diamond)/(I_\uparrow^\diamond + I_\downarrow^\diamond)$, where $I_{\uparrow,\downarrow}^\diamond = \int_{|k_{||}|<2\,\text{Å}^{-1}} I_{\uparrow,\downarrow} dk_{||}$ for the measurements shown in figure 4.9

is due to broadening in the spectral density of the valence electronic structure and not due to inelastic scattering.

4.2.3 Exchange and spin-orbit contribution to the spin polarization

The spin-resolved data in section 4.2.2 was shown for a single polarity 'M+' of the magnetization. Here we present the effect on the spin polarization of photoelectrons after reversing the magnetization while keeping the incidence of light unaltered. The photoemission geometries of our experiment are shown in figure 4.11(a,b).

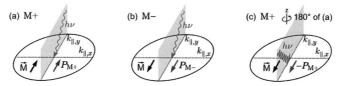

Figure 4.11: (a,b) photoemission geometries of the experiment for magnetization reversal (c) geometry '(a)' after applying $R_{z,180°}$.

Figure 4.11(c) is the equivalent of applying $R_{z,180°}$ to figure 4.11(a), where $R_{z,180°}$ is a rotation around the surface normal by $180°$, This reverses the magnetization and changes the incidence of light, but preserves their relative orientation. The sample surface is symmetric with respect to $R_{z,90°}$, and therefore is in the same state as in figures 4.11(a,b). The value of P in figure (c) is the exact negative of P in figure 4.11(a) as a result of $R_{z,180°}$. However, for the actual experimental situation corresponding to figures 4.11(a,b), $P_{\mathrm{M+}} \neq -P_{\mathrm{M-}}$. We define

$$P_{\mathrm{s\text{-}o}}=(P_{\mathrm{M+}}+P_{\mathrm{M-}})/2 \quad P_{\mathrm{ex}}=(P_{\mathrm{M+}}-P_{\mathrm{M-}})/2 \qquad (4.9)$$

where $P_{\mathrm{s\text{-}o}}=0$ is equivalent to an exact sign change of the spin polarization after magnetization reversal. As shown in section 7.2, $P_{\mathrm{s\text{-}o}}$ and P_{ex} are contributions to the photoelectron spin polarization that can be assigned to spin-orbit coupling and exchange interaction, respectively. An example for the effect of $P_{\mathrm{s\text{-}o}}\neq0$ is presented in figure 4.12, which shows the partial intensity images for spin-up and spin-down and magnetizations 'M+' and 'M-'. If $P_{\mathrm{s\text{-}o}}=0$, then the two images $I_{\uparrow,\mathrm{M+}}(k_{||,x},k_{||,y})$ and $I_{\downarrow,\mathrm{M-}}(k_{||,x},k_{||,y})$ should be equal and contain only majority-spin features, while the images $I_{\downarrow,\mathrm{M+}}(k_{||,x},k_{||,y})$ and $I_{\uparrow,\mathrm{M-}}(k_{||,x},k_{||,y})$ should be equal as well and contain only minority-spin features. We do find the rectangular majority-spin feature P_1 in the majority-spin images for both magnetizations. However, it is also present in the minority-spin image for magnetization 'M-' (figure (c)), while

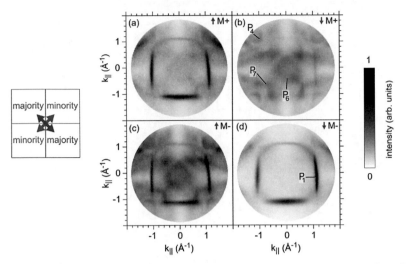

Figure 4.12: spin-up (a,c) and spin-down (b,d) partial intensities measured at $E_i = E_F$ and $h\nu = 50\,\text{eV}$ using p-polarized light.

nearly absent in the minority-spin image for magnetization 'M+'. Similarly, the minority-spin features P_4, P_6, P_7 are absent in the majority-spin image for magnetization 'M-', but stronger enhanced in the majority-spin image for magnetization 'M+'.

The question is, whether this behaviour affects the different resonant photoemission features to the same degree, or not. As mentioned already, this 'asymmetric' behaviour corresponds to a spin-orbit contribution $P_{\text{s-o}} \neq 0$. to the spin polarization. In figure 4.13(a-c), we calculate $P_{\text{s-o}}$ from equation (4.9). The initial result shows $P_{\text{s-o}}$ values averaged across the field of view, which are different from zero and quite large. For the experimental geometry used here (see figure 4.11(a,b)) and p-polarized light, we expect the following property:

$$P_{\text{s-o}}(k_{||,x}, k_{||,y}) = -P_{\text{s-o}}(-k_{||,x}, k_{||,y}) \tag{4.10}$$

The derivation is given in section 7.3. This property is not fulfilled. In section 3.3.4, we discussed that due to an uncertainty in $R_{h,l}$ and $S_{h,l}$ the systematic error of the obtained spin polarization is in the order of 10%. Since we add two independent spin polarization values $P_{\text{M+}}, P_{\text{M-}}$ in equation (4.9), the error of $P_{\text{s-o}}$ is $\approx 14\%$. The homogeneity of $P_{\text{s-o}}$ across the field of view in figures 4.13(a-c) suggests that this error is mainly a constant $u_{P_{\text{s-o}}}$ added to the image $P_{\text{s-o}}^{(m)}(k_{||,x}, k_{||,y})$ (where index (m) means "obtained from the measurements" and (c) means corrected value):

$$P_{\text{s-o}}^{(m)}(k_{||,x}, k_{||,y}) = P_{\text{s-o}}^{(c)} + u_{P_{\text{s-o}}} \tag{4.11}$$

The constant $u_{P_{\text{s-o}}}$ can be removed by requesting that a corrected value $P_{\text{s-o}}^{(c)}$ should have the property given in equation (4.10). The following rule yields such a corrected value

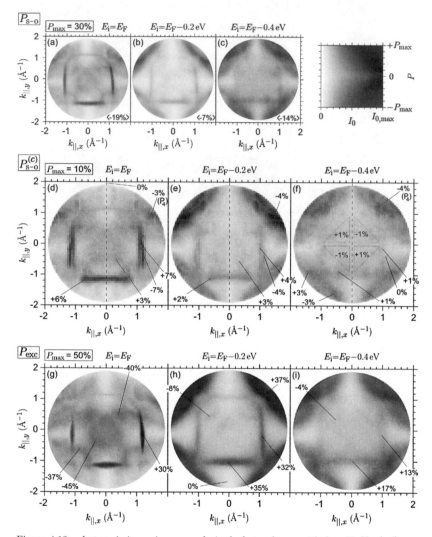

Figure 4.13: photoemission using a p-polarized photon beam with $h\nu=50\,\text{eV}$. (a-f) raw data $P_{\text{s-o}}$ and corrected data $P_{\text{s-o}}^{(c)}$ for the spin-orbit contribution to the photoelectron spin polarization; (g-i) exchange contribution P_{ex} to the photoelectron spin polarization; For reference energy spectra, see figure 4.6. P_{max} refers to the color scale. (a-c) numbers in brackets give the average computed as $\frac{\int P_{\text{s-o}}(k_{||,x},k_{||,y})I_0(k_{||,x},k_{||,y})dk_{||,x}dk_{||,y}}{\int I_0(k_{||,x},k_{||,y})dk_{||,x}dk_{||,y}}$. Further details, see text.

$P_{\text{s-o}}(c)$.

$$P_{\text{s-o}}^{(c)}(k_{||,x}, k_{||,y}) = (P_{\text{s-o}}^{(m)}(k_{||,x}, k_{||,y}) - P_{\text{s-o}}^{(m)}(-k_{||,x}, k_{||,y}))/2 \qquad (4.12)$$

It is designed such that $P_{\text{s-o}}^{(c)} = P_{\text{s-o}}^{(m)}$, if $P_{\text{s-o}}^{(m)}(k_{||,x}, k_{||,y})$ already fulfills equation (4.10), i.e. a correct image of $P_{\text{s-o}}$ is not altered by applying equation (4.12). Using this equation, we obtain the data shown in figures 4.13(d-f). Only the positive $k_{||,x}$ side represents independent data, while the $k_{||,x} < 0$ part, follows from construction. For all data shown here, the absolute values of $P_{\text{s-o}}^{(c)}$ are below 10%, meaning the spin-orbit interaction is small, but measurable. The largest absolute values (± 7) occur for feature P$_1$ at $E_i = E_F$, where we find a 'plus-minus' behaviour. Proceeding to energies below the Fermi level, not only the intensity of P$_1$ decreases, but the spin-orbit effect does, too. At $E_i = E_F - 0.4$ eV, the largest asymmetry is $P_{\text{s-o}} = \pm 4\%$ for feature P$_4$.

4.2.4 valence electronic structure in the WLWL plane

After probing the environment of a mirror plane centered at the Γ point of the fcc Brillouin zone, setting the photon energy to $h\nu = 70$ eV allows for investigating the environment of the L point, as illustrated in figure 4.14(i). Figure 7.4(b) in section 7.4 shows an idealized planar cut through the 4 L points of the upper half of the fct Brillouin zone and the point groups for each wave vector on this plane. The high symmetric points and axes specific to this plane are the L and W points as well as a 2-fold rotational axis connecting L and W point. Additionally, a fourfold rotational axis and two inequivalent mirror planes are present independently of $h\nu$ (figure 7.4(c)). For wave vectors within a mirror plane m, the corresponding Bloch waves are two-fold degenerate and can be represented by wave functions of even (g) and odd (u) parity with respect to m. In conjunction with a suitable polarization and incidence of the photon, either one of these parities can be suppressed. Here, we chose s-polarized light and an azimuth parallel to the y axis. Then, the dipole moment of the photon field is aligned along x such that we can write the electric vector potential in the form $\mathcal{A} = (\mathcal{A}_x, 0, 0)$. Inserting this in the transition matrix element $M_{f,i}$ of equation (2.2), we obtain

$$M_{f,i} \propto \langle f \,|\, \frac{d}{dx} \,|\, i \rangle \qquad (4.13)$$

The inner product $\langle \ldots \rangle$ is an integration over space, i.e. $\iiint \ldots \mathrm{d}x\mathrm{d}y\mathrm{d}z$. The integral vanishes, if the integrand has uneven parity with respect to either x, or y, or z. We know that the final state $f^{(y)}$ of a photoelectron with $\vec{k}_{||} = (0, k_{||,y})$ is invariant under m_x, where m_x transforms $x \rightarrow -x$, and that the operator $\frac{d}{dx}$ changes sign under m_x. Schematically, we can write $M_{f,i} \propto \langle \text{even} \,|\, \text{odd} \,|\, i \rangle$. If the initial state i has even parity with respect to m_x, then the integrand is an uneven function in x and the integral $\int dx$ vanishes. Consequently, initial states with even parity under m_x are forbidden along the y axis in $(k_{||,x}, k_{||,y})$ images. Similarly, one obtains that initial states with odd m_y parity are forbidden along the x axis in $(k_{||,x}, k_{||,y})$ images. m_x and m_y belong to the same class of the symmetry group $C_{4v} = (\frac{4}{m} \frac{2}{m} \frac{2}{m})$, since the fourfold rotation translates one into the other.

In summary, the following selection rules hold: For photoelectrons with azimuth parallel to the y axis, initial states with odd mirror parity are allowed, while along the x

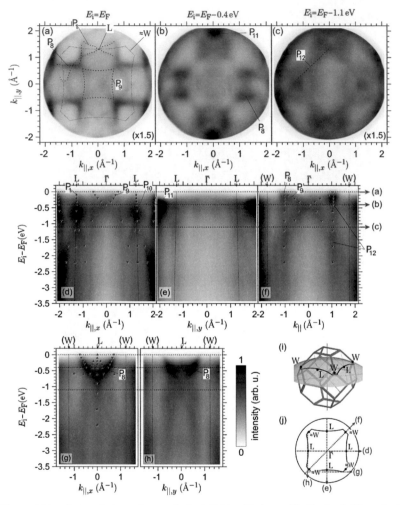

Figure 4.14: photoemission using s-polarized light with $h\nu = 70$ eV. (a-c) $I(k_{\|,x}, k_{\|,y})$ spectra at discrete binding energies; (d-h) $I(k_\|, E)$ spectra with $\vec{k}_\|$ along the directions indicated in (j); (d) even m_x/m_y parity; (e) odd m_x/m_y parity; (h) even m_x/m_y parity at $k_\|=0$; (g) odd m_x/m_y parity at $k_\|=0$; The border of the bulk Brillouin zone is drawn in (a) with dots. P_n are used for reference to photoemission intensity maxima (dashed lines) and to indicate the connection between (a-c) and (d-h). P_1 is connected to the majority part of the Fermi surface ($S_D^\uparrow$), P_8 to $S_B^\downarrow$ (refer to figure 4.1). Red upwards (blue downwards) pointing triangles indicate the predominant electron spin for quasiparticle peaks, as obtained from spin-resolved photoemission at discrete binding energies.

axis, states with even mirror parity are allowed. The mirror parity refers to mirror planes spanned by the surface normal and the (110) direction ($\Gamma - \Sigma - K$, horizontal and vertical axes in our $(k_{||,x}, k_{||,y})$ images). The respective energy spectra are shown in figure 4.14(d) (even parity) and figure 4.14(e) (odd parity). The photoemission feature P_8 has odd mirror parity at the L point, P_{10} has even parity and P_{11} has odd parity. P_9 may have undefined mirror parity. Even m_x/m_y parity is compatible with $\Delta_1, \Delta_2, \Delta_5$, and odd m_x/m_y parity is compatible with $\Delta_1', \Delta_2', \Delta_5$ on the Δ axis [166].

Figures 4.14(a) and 4.15(a) show the Fermi surface cross section obtained from a spin-integral and spin-resolved measurement. P_1 is a contour of the majority-spin part with openings at the L points that connect to neighbouring Brillouin zones. For decreasing binding energies down to $E_F - 0.4$ eV the neck diameter at the L points is observed to shrink as well as the overall area inside P_1. However, for energies $< E_F - 0.4$ eV, the structure is not discernable from other high-intensity features.

P_8, P_9 are a part of the minority-spin Fermi surface, both connected to $S_B^\downarrow$ in figure 4.1. The dispersion of the $k_{||} = 0$-centered feature P_9 is seen in figures 4.14(d,f) while the high-intensity corners P_8 are tracked in figures 4.14(g,h).

The upwards- and downwards pointing triangles inserted in figures 4.14(d-h) indicate the predominant photoelectron spin. We observe, that the high-intensity features are mostly of minority spin. An average exchange splitting of 1.2 eV has been reported [167] and is consistent with photoemission calculations shown in section 5.3, figure 5.6. Hence, the majority spin counterparts of such resonances should appear at lower energies within the recorded energy range. Especially for the highly intense P_8, the majority counterpart should appear in figures 4.14(g,h) at $E_F - 3.2$ eV and above. Another such example is P_{10} in figure 4.14(d). We ascribe the apparent absence of the majority features to a strong energy broadening, which is predicted by DMFT calculations to be ≈ 1 eV for $E_F - 3$ eV, (cf figure 5.3(b)).

Once more, structures are present in the $I(k_{||,x}, k_{||,y})$ spectra, which persist rigidly over large energy intervals, for example the rhombic shape P_{12} in figure 4.14(c), that appears as a vertical bar in the energy spectrum in figure 4.14(f). While the photoemission intensity remains constant down to $E_F - 3.5$ eV, the predominant photoelectron spin does not, as seen for energies $E_F - 1.2$ eV and $E_F - 2.2$ eV. The spin-resolved momentum distributions for these two energies are shown in figures 4.15(f,g). For energies close to the Fermi level, the same trend as for $h\nu = 50$ eV is observed: The Fermi level is dominated by minority states, while the majority spin becomes dominant at $E_F - 0.3$ eV and below – at least when considering the entire field-of-view. In normal emission, the spin polarization is negative for all energies which are shown in figures 4.15(a-g).

4.3 Beyond the bulk valence electronic structure

The focus of this work is the valence electronic region of bulk fcc cobalt. Hence, the interpretation of photoemission spectra was focussed on features which are related to bulk electronic states in the range $E_i \leq E_F$. However, two further types of photoemission features that do not belong to this category, shall be addressed briefly: (1) final-state resonances and (2) surface states/resonances.

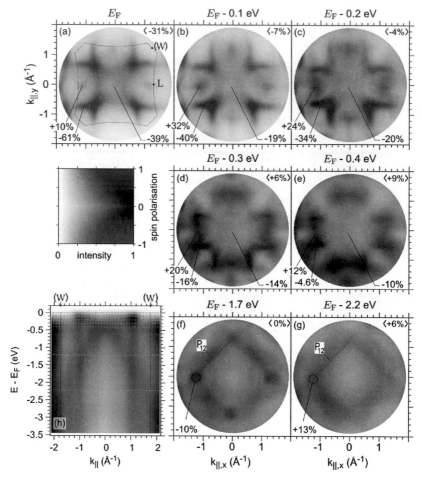

Figure 4.15: (a-g) spin-resolved $I(k_{\|,x}, k_{\|,y})$ photoemission spectra corresponding to excitation with an unpolarized 70 eV photon beam on 10 monolayers of cobalt on Cu(001). Spin polarization values are given for normal emission and further selected points. The average of the spin polarization for the field of view is given in brackets "$\langle \pm n\% \rangle$". (h) $I(k_{\|}, E)$ spectrum as in figure 4.14(f); dashed lines indicate the binding energies of (a-g).

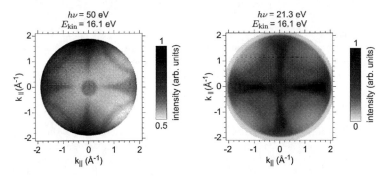

Figure 4.16: Momentum distributions using two different photon energies and the same final-state energy. Independently of the photon energy, the resonant structure has the same shape.

4.3.1 Final-state resonances

Final-state resonances have been characterized in [17] to dominate the photoemission spectra when trying to probe the Fermi surface using photon energies $h\nu=21\,\mathrm{eV}, 45\,\mathrm{eV}$. Our observations contradict this statement. A characteristic of final-state related photoemission peaks is that they occur at a fixed final state energy E_f independent of the photon energy $h\nu$. This behaviour we find in figure 4.16, where we have selected the same $E_f=E_{\mathrm{vac}}+16.1\,\mathrm{eV}$ and used two photon energies $h\nu=21.3\,\mathrm{eV}, 50\,\mathrm{eV}$. The enhanced-intensity regions form the same shapes for both photon energies, a cross like shape, a filled circle centered around normal emission and arcs connecting the K points. Assuming a work function $\phi=4.7\,\mathrm{eV}$ of the cobalt surface, this final state resonance affects measurements probing the Fermi surface with $h\nu=20.8\,\mathrm{eV}$, while for He-I radiation, this final-state resonance appears in a binding energy scan at $E_i - E_{\mathrm{F}}=-0.5\,\mathrm{eV}$. No evidence of further final-state resonances has been found in the photon energy range $35\,\mathrm{eV}<h\nu<200\,\mathrm{eV}$ (cf. figure 4.2). The conclusions in [17] may be drawn because of insufficient intensities. The high-intensity final-state resonance and the intensity drop towards higher energies may have been mistaken for the intensity cutoff at the Fermi level.

4.3.2 Surface resonances

A minority surface resonance in Co/Cu(001) has been mentioned in several publications [21, 168, 169] located at $E_{\mathrm{SR},\downarrow}=E_{\mathrm{F}}-0.4\,\mathrm{eV}$ ($E_{\mathrm{F}}-0.45\,\mathrm{eV}$) for $k_{\parallel} = 0$. Its dispersion along the $\bar{\Gamma}\bar{X}$ axis of the surface Brillouin zone has been obtained in [168] by comparing energy spectra at discrete angles for a clean and a contaminated surface.

For the interpretation of momentum-resolved photoemission experiments, it is desirable to compare to theoretical data that shows the spectral distribution of surface states and resonances for the *entire surface Brillouin zone (SBZ)*. The layer- and spin-resolved spectral density presented here has been calculated using the omni code [147]. Surface resonances (SR) and surface states are characterized by an enhanced probability density at the surface layer compared to the bulk. In order to quantify and visualize this enhancement, we plot the asymmetry between the surface layer and an average of the layers 2 to 10 for majority and minority electrons in figure 4.17. The calculation does not include energy renormalization as caused by electron-electron interaction (section 5.3 is dedicated to this topic). Hence, the energy axis is not absolute and quasiparticles may be energetically shifted with respect to experimental data. This shift may depend on wave vector, binding energy, spin, localization, etc. In figure 4.17, we present selected results for the minority-spin. The patterns in figure 4.17(c) are similar to the $I_\downarrow(k_{||,x}, k_{||,y})$ minority-spin spectrum in figure 4.9(a) recorded for initial states at the Fermi level. In particular, the k-diameter of $SR_{1\downarrow}$ matches the outer ring in figure 4.6(a). Based on these corresponding features, the energy axis of figure 4.17(a,b) was calibrated. Surface resonances for $k_{||}=0$ then occur at energies $E = E_F - 0.1$ eV and $E = E_F - 1.0$ eV, in contrast with the experimental results previously published. We attribute this discrepancy to renormalization effects. Consequently, the experimental dispersion from [168] shown in figure 4.17(a) as solid squares, does not agree. Furthermore, no similar dispersive surface resonance is present within the energy range of our calculation. The agreement may improve partially, if $SR_{0\downarrow}$ and $SR_{2\downarrow}$ are shifted to higher energies and the data points [168] are ascribed to these two separate surface resonances. A study of clean and contaminated fcc Co surfaces with momentum-resolved photoemission may give more insight. The result within this work is that the surface resonance marked as $SR_{1\downarrow}$ in figure 4.17(c) most likely corresponds to the feature marked as "SR?" in figure 4.6(a).

4.4 Unoccupied electronic states probed by two-photon photoemission

4.4.1 Quantum well states in Co/Cu(001)

Now we turn to experiments conducted in our home laboratory, where we probed the unoccupied electronic structure of Co thin films on Cu(001). To this purpose, two-photon photoemission measurements were performed using ultra-short laser pulses. These were generated in a home-built mode-locked Ti:sapphire oscillator with subsequent frequency doubling in a BaB_2O_4 crystal (BBO). The BBO crystal outputs light pulses with a central photon energy of 3.07 eV [170]. The pulse repetition rate was 80 MHz corresponding to a 1.9 m length of the cavity. The duration of one pulse is in the order of 20 fs and the average output power ranges from 80 mW to 120 mW corresponding to pulse energies of 1 nJ to 1.5 nJ or $2 \dots 3 \times 10^9$ photons per pulse.

In our case, two photons of the energy 3.05±0.05 eV are absorbed. Therefore, intermediate states can be probed in a region $E_F + 1.6$ eV $< E < E_F + 3.1$ eV. The work function for the fct Co thin film $\phi = 4.8 \pm 0.2$ eV follows from our experiments and is in accordance with [171]. For lower intermediate state energies, no photoelectron is generated, whereas for higher intermediate state energies, the corresponding initial state has an energy above the Fermi level and is not occupied by an electron. The alignment of band gaps at the Co/Cu(001) interface which determines the energy range of quantum well states was shown in figure 2.4. In bulk Cu(001), the electronic states along the Δ direction, which

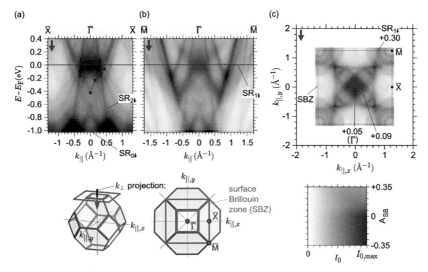

Figure 4.17:   Enhancement of the layer-resolved *minority-spin* spectral density $A_l(E, k_{\|,x}, k_{\|,y})$ in the surface layer $l=1$ with respect to the average $\bar{A}(E, k_{\|,x}, k_{\|,y})$ of the layers $2 \leq l \leq 10$. The asymmetry $A_{\mathrm{SB}} = \frac{A_1 - \bar{A}}{A_1 + \bar{A}}$ characterizes electronic states as surface resonances/states ($0 < A_{\mathrm{SB}} < 1$), or bulk states ($-1 \leq A_{\mathrm{SB}} \leq 0$). For the energy interval shown here, $A_{\mathrm{SB,max}} \lesssim +0.35$, while for image-potential states $A_{\mathrm{SB,IP}} \gtrsim +0.80$. The data has been calculated for the (001) surface of 12 monolayers Co/Cu(001) using **omni** [147]. (a,b) Energy-$k_\|$ maps for (a) the $\bar{\Gamma}\bar{X}$ direction, (b) the $\bar{\Gamma}\bar{M}$ direction, (c) $k_{\|,x}$-$k_{\|,y}$ map for the nominal energy $E_{\mathrm{F}}+0.6\,\mathrm{eV}$. Solid squares in (a) are data from [168].

are probed in normal photoemission on a Cu(001) surface, extend from the valence band region up to an energy $E_F + 2.0$ eV according to band structure calculations in [102, 172] or up to $E_F + 1.6$ eV in [8], followed by a gap whose upper bound is at $E_F + 8.0$ eV. The occurence of quantum well states is restricted to the energy region where the band gap in Cu(001) coincides with the availability of electronic states in the Co overlayer along the Δ direction. These electronic states are derived from a spin-split band of Δ_1 symmetry that extends up to $E - E_F = 3$ eV (minority) or 2.8 eV (majority electrons) [8] (data reproduced in figure 4.1(e)).

Figure 4.18(a-d) shows the parallel-momentum distribution of photoelectrons for a 6 monolayer Co thin film on Cu(001) for selected kinetic energies that correspond to initial and intermediate state energies as labelled above the graphs. The sample temperature was 300 K. The azimuth of the incidence of light is shown in figure (a). For this film thickness d, we see exactly one order of the quantum well state. Its associated $k_\perp^{QW}$ component as well as its energy $E(k_\perp^{QW})$ are a function of d, as follows from the phase accumulation model (section 2.7). $k_\perp$ is within the red bar in figure 4.18(g). In figure 4.18(a), the resonant two-photon process is observed at $k_\parallel = 0$, while it moves to larger $k_\parallel$ radii when we select higher energetic photoelectrons (b-d). The interpretation in terms of the three-dimensional band structure is shown in figure 4.18(f): The quantum well state disperses to higher energies, when deviating from the symmetry axis Δ. Scanning the energy in 50 meV steps and extracting the intensities along the vertical ($k_{\parallel,y}$) axis, we extract the in-plane dispersion shown in figure 4.18(e). Here, $k_{\parallel,y}$ corresponds to a bulk (100) direction which is indicated by the small drawing in the lower right corner of the graph. In the surface parallel directions, $E(k_{\parallel,x}, k_{\parallel,y})$ is a continuous function, since there is no confinement to quantum-size scales. Therefore, this $k_\parallel$-dispersion is equivalent to the dispersion in a bulk (fcc) Co crystal. One limitation of this statement shall be mentioned: The $k_\perp$ component of the quantum well state is a function of $k_\parallel$. The reason can be given by the phase accumulation model: the value of $k_\perp$ which solves equation (2.31), depends on the phase change ϕ_C during reflection at the substrate-overlayer interface. ϕ_C, in turn, depends on the lower and upper limits of the band gap in the substrate. This band gap is a function of the surface-parallel components of the wave vector $(k_{\parallel,x}, k_{\parallel,y})$. The consequence is, that the dispersion relation in figure 4.18(e) is not obtained along a straight line on the planar orange surface in figure 4.18(f). Rather, the set of quantum well state wave vectors probed in the energy scan form a curved plane. Such a $k_\perp(k_\parallel)$ dependence has been shown in [173] for quantum well states in copper (Cu/Co/Cu(001)).

In figure 4.18(b,c), an intensity asymmetry occurs between the left and right part. For the measurements shown here, p-polarized light was used incident from the horizontal direction with a polar angle of $25°$. If the electric field of the photons is decomposed to the crystallographic axes, one has a surface normal component and a surface-inplane component along the horizontal direction. The latter causes the left-right asymmetry in the momentum distribution. Measurements with s-polarized light show an asymmetry between the top and bottom half of the image, accordingly.

The close-to-circular momentum pattern in figures (a-d) tells us, that the dispersion in the $k_\parallel$-plane is rather isotropic[2]. Due to the four-fold rotational symmetry of the

---

[2]Photoelectrons along the diagonals correspond to different directions in the bulk crystal than the

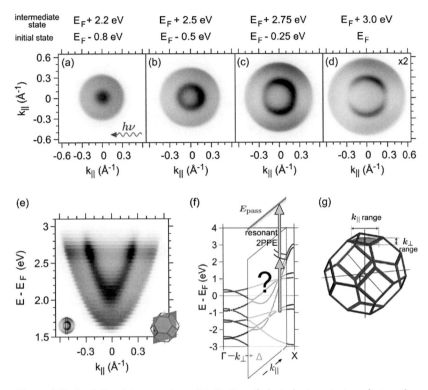

Figure 4.18: (a-d) Parallel-momentum distribution of photoelectrons in two-photon photoemission ($h\nu = 3.05$ eV, p-polarized) for a 6 monolayer Co thin film on Cu(001). Several binding energies are shown from left to right. The intensity shown in the right-most graph has been multiplied by 2. (e) $E(k_{\parallel,y})$ map of the photocurrent reveals a parabolic (in-plane) dispersion of the quantum well state. (f) sketch of the resonant (step-by-step) two-photon process via the unoccupied fcc Co bands in the bulk limit. For the 6-monolayer thin film, quantum well states with discrete values of $k_\perp$ and $E(k_\perp)$ replace this band. Thereby, the $k_\perp$ selection is determined by the film thickness. For a selected photoelectron energy, the resonant transition may occur for $k_\parallel \neq 0$, as in the depicted example. (g) The region of the Brillouin zone that is probed by the resonant two-photon transitions via quantum well states. The $k_\parallel$ field of view in (a-d) is a projection of the orange surface. The $k_\perp$ range is the result of mapping the allowed energy range to the $k_\perp$ axis by the dispersion of the calculated unoccupied bulk band.

cubic (001) surface, the peak positions along the horizontal and vertical axes should be equidistant from the center ($k_{||} = 0$) in figures 4.18(a-d). They are not in figure (c). In figure (d), the intensity along the horizontal axis is diminished. These anisotropies are due to the off-normally incident p-polarized light which imposes different selection rules for photoelectrons emitted along the vertical or horizontal axis. This can affect both excitation steps (pump and probe) of the resonant two-photon process.

4.4.2 Dispersion and thickness dependence

Here, we show how the quantum well state dispersion changes for varying thickness d of the Co overlayer. For each thickness, a new Co thin film was grown on a freshly prepared Cu(001) substrate (2 hours sputtering with 2 keV Ar$^+$ ions, annealing at 600°C for 5 min). Figure 4.19 shows the energy distribution curves resolved for a $k_{||}$ component along the bulk (100) direction ($\bar{\Gamma}\bar{M}$ in the (100) surface Brillouin zone) for increasing thickness of the Co overlayer. The enhanced photocurrent parabolic feature caused by resonant 2PPE via the quantum well states shifts to higher energies with increasing thickness. For 11 monolayers of Co, a second order ν of quantum well state appears at $E_F + 1.9$ eV and also shifts to higher energy for the 13 monolayer film (see figure 2.5 for a visualization of the different orders $\nu = 1, 2$). For selected energies, intensity line profiles have been plotted to determine the $k_{||}$ values of the peaks. These are shown as filled blue and red circles. The grazing emission threshold was fitted for each energy by an erf($k_{||}$) function and the central $k_{||}$ value is shown as green circles. In both cases, the resulting $E(k_{||})$ dispersion has been fitted by a parabola. For the grazing emission threshold, this is the exact description since at a given kinetic energy of the photoelectron travelling to the detector, the maximum $p_{||} = \hbar k_{||}$ is given by $E_{\text{kin}} = p_{||,max}^2/(2m_e)$. For the $E(k_{||})$-dispersion of the quantum well state, the parabolic fit is a simple approximation which enables us to quantify the actual dispersion with two parameters: $E_0 = E(k_{||} = 0)$ and m_{eff}. The effective mass is defined as $m_{\text{eff}} = \hbar^2(\partial^2 E/\partial k^2)$, i.e. the inverse of the curvature a of an $E(k)$ dispersion. We exploit the grazing photoemission threshold as a precise gauge for the $k_{||}$ axis, since its curvature a_0 yields the "effective mass" of the free electron. The effective mass parameter in units of the free electron mass is then obtained by the ratio a_0/a_1, where a_1 was the curvature of the quantum well state. The dependence of E_0 and m_{eff} on the film thickness is summarized in table 4.2 and plotted in figure 4.19(b). The monotonously rising energy as a function of thickness is due to the the Δ_1 band, from which the quantum well states are derived. It has a maximum at the X point. This dispersion is probed by the $k_\perp(d)$ which moves towards the X point as a function of the film thickness.

Usually, if data for a wider range of d exists, the value of $k_\perp$ can be calculated $k_\perp = \frac{\pi(\nu_2 - \nu_1)}{d_2 - d_1}$ with thickness values d_1 and d_2, where the orders ν_1, ν_2 of the quantum well states have the same energy values $E_{\nu_1} = E_{\nu_2}$ (follows from equation 2.31). In that way, the band dispersion can be obtained from experiment [116]. With the kind of $k_{||}$-resolved photoemission data as shown in figures 4.19, the function $k_\perp(k_{||})$ can be obtained. In [173] this dependence was shown to have an influence on the experimentally assessed effective mass of the $E(k_{||})$ dispersion. Therefore, the effective mass parameters given in table 4.2

---

horizontal and vertical axes. Therefore the peak positions along the diagonal directions may in principle have a different k-distance to $k_{||} = 0$.

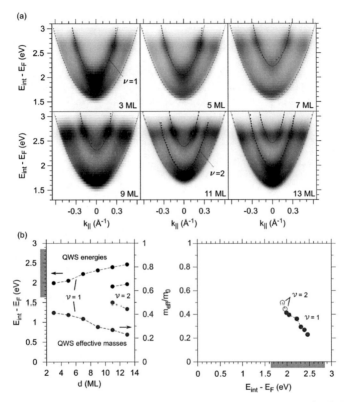

Figure 4.19: (a) $E(k_{\parallel})$ photoemission intensity maps with $k_{\parallel}$ along the bulk (100) direction for Co film thicknesses from 3 to 13 monolayers. (b) left graph: Energies of the quantum well states (intermediate state energy) in normal photoemission and fitted curvature of the $E(k_{\parallel})$-dispersion expressed as effective mass; right graph: effective mass fit of the $E(k_{\parallel})$-dispersion as a function of the quantum well state energy at $k_{\parallel} = 0$

d (ML)	$E_{QW} - E_F$(eV)	m_{eff}	$E_{QW} - E_F$(eV)	m_{eff}
3	1.99	0.413	-	-
5	2.05	0.395	-	-
7	2.22	0.362	-	-
9	2.31	0.293	-	-
11	2.39	0.268	1.90	0.499
13	2.46	0.229	1.96	0.444

Table 4.2: Energies of the quantum well states (intermediate state energy) in normal photoemission and fitted curvature of the $E(k_{||})$-dispersion expressed as effective mass

may require a small correction. Due to the strong dispersion of the Δ_1 bulk band from which the quantum well states are derived, the relevant $k_\perp$ range is rather small and the correction due to $k_\perp(k_{||}) \neq$ const does not significantly alter the values.

4.4.3 Non-resonant two-photon processes versus one-photon processes

In addition to the quantum well state resonances seen in two-photon photoemission, one further enhanced-intensity structure is distinct in figure 4.19: a horizontal bar at $E_F + 2.7$ eV, corresponding to an initial state energy $E_i = -0.4$ eV. It does not shift in energy with varying thickness of the cobalt thin film. An additional experiment has been performed to unambiguously assign this feature to the occupied valence electronic structure, i.e. electronic states at $E_F - 0.4$ eV. By doubling the frequency of the laser pulses with a second β-Ba$_2$BO$_4$ crystal, we obtain photons with 6.05 eV (highest energy that can be produced by second harmonic generation [174] with this crystal). Since this is close to the value of the sum energy of our two-photon photoemission measurements ($2h\nu_{2PPE} = 2 \times 3.1$ eV), the final and initial states involved in both non-resonant (direct) two-photon transitions as well as one-photon transitions ($h\nu = 6.05$ eV) are rather similar. However, the 1PPE measurement is not sensitive to states at $\approx E_F + 3.1$ eV. The comparison between 2PPE and 1PPE is shown in figure 4.20. The $E(k_{||})$-map of both 1PPE and 2PPE show a distinct enhanced-intensity structure at an initial state energy of $E_F - 0.4$ eV and $k_{||} = \pm 0.45 Å^{-1}$. In the $(k_{||,x}, k_{||,y})$ map, the same feature appears as a rather undefined shape extended over the complete $k_{||}$-radius of $0.4 Å^{-1}$, but its intensity is weaker along the horizontal axis than along the vertical axis. It is again found for both 1PPE and 2PPE measurements. A second feature visible in 1PPE that connects the points $(E_1, k_{||,1}) = (E_F - 0.5$ eV$, 0)$ and $(E_2, k_{||,2}) = (E_F - 0.1$ eV$, 0.35 Å^{-1})$ is not found in 2PPE.

4.4.4 Spin polarization in resonant two-photon photoemission

So far, we have shown $(k_{||,x}, k_{||,y})$ maps as well as the in-plane dispersion in $E(k_{||})$ maps of quantum well states that are derived from the unoccupied Δ_1 bulk band of cobalt close to the X point, at $E_F + 2.8$ eV. DFT+LSDA theory predicts that this band is spin-split (fully relativistic calculation includes both exchange interaction and spin-orbit coupling) by 0.2 eV at the X point. Our energy resolution of 0.1 eV is sufficient to resolve that splitting. Contrarily, in figures 4.19 and 4.18, we observe a single enhanced intensity

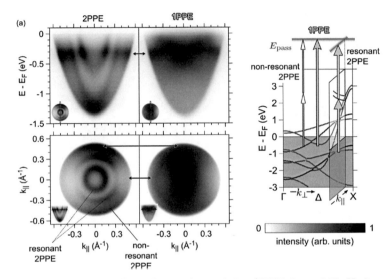

Figure 4.20: comparison of one-photon photoemission (1PPE, $h\nu = 6.05$ eV, 6 mono-layer Co/Cu(001)) to two-photon photoemission (2PPE, $2h\nu = 2 \times 3.1$ eV, 11 monolayers Co/Cu(001)); Both measurements used p-polarized light. Arrows emphasize the similari-ties between 1PPE and 2PPE. The figure to the right shows that for resonant two-photon photoemission, we are limited by the unoccupied Δ_1 band to a $k_\perp$ range close to the X point. Contrarily, non-resonant (direct) 2PPE transitions as well as 1PPE transitions are in principle possible for the whole $k_\perp$ range and so probe different regions of the bulk Brillouin zone. Therefore, the states which are observed in 1PPE are not necessarily the initial states of the resonant (step-by-step) 2PPE process via the QWS.

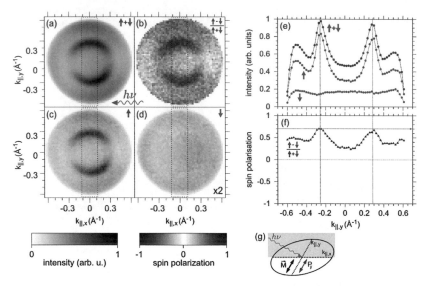

Figure 4.21: spin-resolved two-photon photoemission of a 6 monolayer Co thin film on Cu(001) probing intermediate states at $E_F + 3.0$ eV (initial states at E_F) corresponding to figure 4.18(d); (a) total photocurrent, (b) spin polarization, (c) majority and (d) minority partial intensities; the displayed intensity in (d) is multiplied by 2; (e-f) line profiles along the vertical axis in (a-d)

feature, or two features with a distinctly larger difference in energy due to different orders $\nu = 1, 2$ of quantum well states. Performing a spin-resolved experiment, we can distinguish whether the single QWS-related intensity feature is produced by a superposition from both spins of the QWS or whether just one of the majority or minority spins is contributing.

Figure 4.21 presents the spin-resolved results where the sample and illumination parameters correspond to the spin-integral measurement shown in 4.18(d). The ring-like intensity feature connected to the step-by-step two-photon process via the QWS is found to have increased spin polarization (figure 4.21(b,f)) with a maximum value of $+70\%$, while the spin-polarization of the non-resonant transitions ranges from $+25 \ldots +50\%$. The sign '+' corresponds to majority spin and the correspondence to the sign of the measured intensity asymmetry was identified by comparison to previous measurements, where the dominant spin character was known. In figure 4.21(d,e), the minority spin partial intensity is almost constant within in the $(k_{\|,x}, k_{\|,y})$-map and shows no sign of the QWS feature. From the theoretical band structure (figure 4.1(e)), and the sketch of the in-plane dispersion in figure 4.18 it is to be expected, that the minority QWS feature appears at a smaller $k_\|$-radius (inside the majority QWS). This is within the field of view in figure 4.21(d), but no enhanced minority intensity is found. Figure 4.22 shows the spin polarization for several binding energies. The spin polarization of the photoelectrons that have been excited via the majority QWS reaches maximum values in the range of 70% to 80% for all three binding energies. None of the spin-polarization $(k_{\|,x}, k_{\|,y})$-maps show a

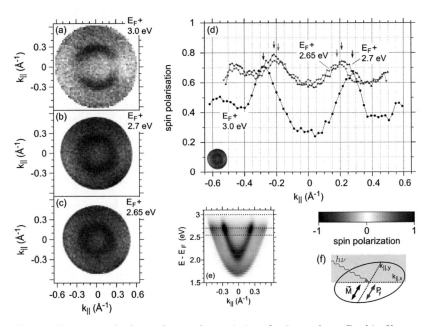

Figure 4.22: spin-resolved two-photon photoemission of a 6 monolayer Co thin film on Cu(001), (a) total photocurrent, (b) spin polarization, (c) majority and (d) minority partial intensities

trend towards minority spin within the area enclosed by the majority QWS feature, except for figure 4.22(a). However, for that energy, the minority partial intensity in figure 4.21 showed no intensity feature connected to the minority QWS. The spin polarization of the photoelectrons excited by non-resonant two-photon processes decreases towards higher energies. This can be reasoned by the spin polarization of the initial states: figure 4.15 shows that the minority spectral density predominates at E_F, but the sign of the average spin polarization reverses at about $E_F - 0.3\,\mathrm{eV}$. The initial state spin polarization may account for the decreasing spin polarization from figures 4.22(c) to (a). However, the predominating majority spin for non-resonant (direct) 2PPE processes in figure (a) demonstrates that the non-resonant 2PPE processes do not lead to the same spin polarization of photoelectrons as one-photon photoemission with $h\nu = 6.1\,\mathrm{eV}$ which directly probes initial states. Possibly, the intermediate state (the majority QWS) participates in the non-resonant 2PPE processes, altering the selection of the electron spin.

4.4.5   Spin-orbit hybridization of majority and minority QWS

All spin-resolved 2PPE experiments shown up to now have been performed on a homogeneously magnetized cobalt thin film with the magnetization pointing in the same direction. In figure 4.23, we compare the majority spin and minority spin intensities of photoelectrons for opposite magnetization directions. The photoemission geometry is the same as shown in figure 4.22(f). We have put the experimental spin-up channel for magnetization M+ in the same row as the spin-down channel for M- such that majority spin is displayed in the first row (figures (a,b) and minority spin in the second row (c,d). For the hypothetical case that all photoemission effects related to spin-orbit coupling are absent and we probe a purely exchange-split electronic structure, we would expect the spin-up of channel M+ to be equal to the spin-down channel of M-, i.e. photoemission intensities in majority spin and minority spin channel each are independent of the sign of the magnetization. A contrasting situation is found most clearly when comparing figures 4.23(c) and (d). In the latter graph, the photoemission intensity within the circular ring that corresponds to resonant two-photon processes via the QWS is also enhanced with respect to the non-resonant "background". However, the k-radius of this feature does not deviate perceivably when comparing the $(k_{||,x}, k_{||,y})$-maps for reversed magnetization. In the line profile (figure 4.23(e)), we observe that for M- the QWS peak intensity in the majority spin channel is partially redistributed to the minority spin channel with respect to M+. A second observation is that for M-, the QWS peak positions of majority spin channel and minority spin channel are shifted by 0.02Å^{-1} with respect to each other. While this is a small effect, we obtain the same shift for $k_{||,y} > 0$ and $k_{||,y} < 0$ [3].

Analogous to section 4.2.3, the exchange and spin-orbit contribution P_{ex} and $P_{s\text{-}o}$ to the spin polarization have been calculated by equation (4.9) and are shown in figures 4.23(f,g), respectively. Their lineprofiles are given in figure (h). The exchange part has the meaning

---

[3]In this work it is always ensured that majority and minority channels were recorded with the exact same projection settings of the electron image in the momentum microscope as well as under the same condition of the sample. This is achieved by periodically switching between the two working points l and h of the spin detector during the spin-resolved measurement, which are used to obtain the images of P and I_0. Inconsistencies in the $k_{||}$-axes are thus excluded, as well as aging effects. The data for M+ compared to M- do not have the same level of consistency, since these are recorded in subsequent experiments.

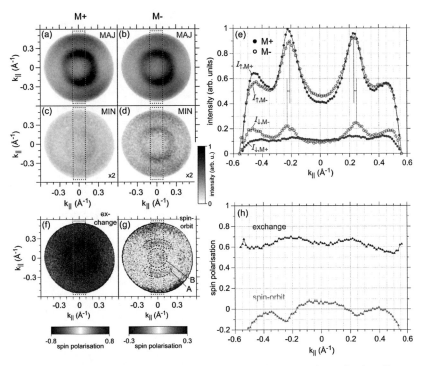

Figure 4.23: spin-resolved two-photon photoemission of a 6 monolayer Co thin film on Cu(001) with magnetization reversal (probing $E_F + 2.7\,\mathrm{eV}$), (a,b) majority (c,d) minority partial intensities; (a,c) original (b,d) reversed magnetization; (e) lineprofiles along the vertical direction as indicated in (a-d) corresponding to the (110) bulk direction; Magnetization and spin-quantization axis are perpendicular to the azimuth of the photon incidence. (f) exchange and (g) spin-orbit contribution to the spin polarization; (h) lineprofiles of (f) and (g)

of a spin polarization that reverses its sign upon magnetisation reversal. The spin-orbit part can be interpreted as follows: Where $P_{s\text{-}o}$ is positive, we detect more majority (or less minority) electrons than would be detected if spin-orbit coupling was absent. Figure 4.23(g) shows, that in the regions marked "A", minority electrons are added, while in the center region marked "B", majority electrons are added. Due to spin-orbit hybridization, the quantum well states are not fully spin polarized. Rather, the majority QWS overlaps with a (relatively small) minority spin spectral density, and vice versa. To what degree the "antagonistic"-spin electrons are excited, depends on the relative orientation of magnetisation and photon incidence (which changes when we reverse the magnetisation, since our light source is fixed). Hence, for M-, we detect a smaller minority-spin peak (curve $I_{\downarrow,M-}$ in figure 4.23(e)) at the position of the majority QWS, which is absent for M+ (curve $I_{\downarrow,M+}$ is nearly flat). This can be interpreted as a small "mixed-in" degree of minority spin to the majority QWS peak. Likewise, the previously-mentioned inwards-shift of the peaks in figure 4.23(e) ($I_{\uparrow,M-}$ with respect to $I_{\downarrow,M-}$), may be interpreted as a small majority-spin contribution to the otherwise hidden minority QWS peak. This presumes that the actual minority QWS corresponds to a smaller k-radius than the majority QWS.

A photoemission calculation (with electron correlation effects turned off) illustrates the position of the minority-spin QWS within the $k_{||}$-plane. Figure 4.24 has been calculated as a one-photon photoemission process with $h\nu = 3.07\,\mathrm{eV}$ and initial state energies in the unoccupied part of the fct Co electronic structure, as if those states were occupied by electrons. In the calculation, a constant lifetime broadening of $0.1\,\mathrm{eV}$ $(1.0\,\mathrm{eV})$ was used for states below (above) the vacuum level. These values correspond to a weakly interacting electron ensemble. Specifically, both spins have the same lifetime. In comparison to our 2PPE experiments, this calculation corresponds to the second step of the resonant (step-by-step) two-photon processes. In the experiment, the pumping of electrons to the minority quantum well state may be suppressed, either because no minority electrons are available at the right energy and wave vector, or because of wave function symmetries that prevent an excitation (forbidden transitions). These mechanisms which may hide the minority quantum well state in 2PPE, are artificially turned off, here.

The calculation confirms, that the minority QWS has a smaller k-radius than the majority QWS and that the energetic position is such, that it should appear in our spectra, provided that the pumping step of the two-photon process provides the minority electrons. The absence is further discussed in the following section.

A schematic in figure 4.25(a) shows the analogue of the calculated energy spectrum (figure 4.24(d)) where the imaginary part of the self energy for the minority electrons is a factor two larger than for the majority electrons. Resuming the discussion of spin-orbit coupling effects, figure 4.25(b) shows a schematic $k_{||}$-spectrum of the QWS spectral density corresponding to figure 4.23(e). The broadening of the minority QWS leads to a significant overlap with the majority QWS spectral density. The consequence is that the "QWS↑" peak in figure 4.25 attains an additional minority-spin intensity contribution for M+, and vice versa for the "QWS↓" peak and M-.

Thus, in the spectrum $I_{\downarrow,M-}(k_{||})$ we observe a minority-spin peak at the position of the majority QWS peak. More interestingly, the transfer of spectral weight to smaller k-radii

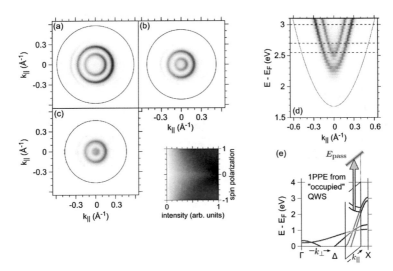

Figure 4.24: calculation of one-photon photoemission with $h\nu = 3.07\,\text{eV}$ and p-polarized light for a 6 monolayer Co thin film on Cu(001). Photoemission initial states above Fermi level are occupied.

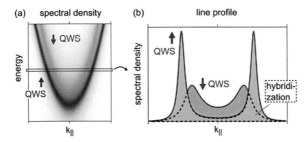

Figure 4.25: (a) Scheme of figure 4.24 with broadened minority-spin quantum well state; (b) Scheme of the spectral density corresponding to figure 4.23(e).

between curves $I_{\uparrow,M+}$ and $I_{\uparrow,M-}$ may be a sign of the minority QWS peaks, of which we observe a small majority-spin signal due to spin hybridization.

4.4.6 Conclusion

Both the spin-integral experiments for different thicknesses of Co thin films as well as spin-resolved measurements have not directly revealed the minority quantum well states, although their presence has been proven by spin-resolved inverse photoemission experiments [19]. This is most likely due to a combination of two effects:

(1) The lifetime of the minority quantum well state is shorter than the majority state, such that its spectral weight is distributed over a larger volume of k-space and energy. Thus, we expect a broadened feature with low intensity. Lifetimes of "hot" (excited) electrons have been studied by time- and spin-resolved two-photon photoemission[175]. Minority electrons 1 eV above the Fermi level were observed to live half as long as their majority counterparts. This ratio is increasing towards higher energies. Hence, we expect the minority quasiparticles (at $\approx E_F + 3\,\mathrm{eV}$) to have less than half the peak intensity (and more than double the peak width) of the majority quasiparticles. A calculation of hot electron lifetimes for Fe shows minority-spin lifetimes five times shorter than for the majority spin [176], corresponding to an energetic broadening of 1 eV. For this order of magnitude, the minority QWS would be barely discernible in our photoemission spectra.

(2) The majority spectral density below the Fermi level is much stronger broadened than the minority spectral density. Therefore, direct transitions of majority electrons to intermediate states in the first excitation step of the two-photon process are less selective in $k_\parallel, k_\perp$ and energy than for minority electrons. Hence, the resonant two-photon processes via the minority QWS may be possible only for particular wave vectors and energies. As argued in [21], the minority surface resonance at $E_F - 0.4\,\mathrm{eV}$ has small spatial overlap with the QWS and therefore is negligible as a pumping reservoir for the minority QWS. In order to populate the minority QWS, minority-spin bulk states are required in the vicinity of the X point for energies $E_F \ldots E_F - 1\,\mathrm{eV}$.

The only minority-spin states in the vicinity of the X point belong to a band with Δ_1 symmetry along the Δ axis ("P_4" in figures 4.6(a,e)). This implies even parity along all mirror planes of the fcc (001) surface. The quantum well states probed by our 2PPE experiments have the same symmetry. Evaluating the dipole selection rules, the highest transition probability occurs along the $k_{\parallel,x}$ axis[4].

Other than in previously published results of *normal* photoemission, here these initial states may provide minority electrons to be excited to the minority QWS. However, as can be seen in figure 4.6(e), the band has a large slope $dE/dk_\parallel$, i.e. a low spectral density per energy interval and therefore a low transition probability. Furthermore, direct energy- and wave-vector-conserving transitions to the minority QWS may indeed be rare.

---

[4]The photon electric field components point in the horizontal ($k_{\parallel,x}$) and surface-normal directions. Along the $k_{\parallel,y}$ axis, the transition matrix element for the first step of the 2PPE process then combines two terms with even parity (initial and QWS state) and one term with odd parity (components of the electric dipole of the photon). Therefore, direct transitions are forbidden for the x component of the photon electric field. For wave vectors on the $X - U - W$ plane (border of the Brillouin zone), direct transitions excited by the surface-normal component (z) are forbidden, as well. Since the QWS are close to the X point, the z component may produce a low transition rate.

In summary, the apparent absence of the minority QWS in two-photon photoemission that contrasts the easily observable majority QWS is consistent with strong broadening of majority-spin states below the Fermi level and of minority-spin states above the Fermi level, both due to electron-electron interaction. We have observed a sizeable effect of spin-orbit coupling by changing the relative orientation of magnetization and photon incidence in spin-resolved two-photon photoemission. Possibly, the spin-orbit hybridization is enhanced by a broader distribution of majority and minority spectral weight which enhances the overlap on the energy axis and in k-space.

Chapter 5

Discussion

5.1 The complex band structure

Electrons in infinitely extended crystalline solids can be described as Bloch waves, i.e. wave functions of the form

$$\phi(\vec{r}) = e^{i\vec{k}\vec{r}} u_{\vec{k}}(\vec{r}) \tag{5.1}$$

where $u_{\vec{k}}(\vec{r})$ has the same periodicity as the crystal lattice. The bulk electronic structure is then the ensemble of real energy eigenvalues (corresponding to stationary states) $E(\vec{k})$ ($\Im(E) = 0$) restricted to the domain of real-valued wave vectors ($\Im(\vec{k}) = 0$). However, in general, further real-valued energy eigenvalues can be found when complex-valued wave vectors $\vec{k} = \vec{k}_r + i\vec{k}_i$, are allowed. Inserting a complex $\vec{k}$ in equation (5.1) means, that the amplitude of the associated wave functions (equation 5.1) grows/decays exponentially along the direction given by $\vec{k}_i$. Such solutions are known to contribute to surface state wave functions [177], namely for the depth regime below the surface, where the electronic potential is indistinguishable from the bulk potential (see shaded area in figure 5.1(e)). We continue the discussion in a one-dimensional subspace of the reciprocal lattice. According to an argument by Heine [178], a line of imaginary wave vectors is allowed (in the sense of *real* energy eigenvalues) in the vicinity of every extremum $dE_n/dk_r|_{k_r=k_{r,0}} = 0$ that occurs in the bulk electronic bands $E_n(k_r)$ along the real wave vector axis. In that case, $E(k) \approx E_0 + (\text{real constant})(k - k_{r,0})^2$. Inserting $k = k_{r,0} + i\,k_i$, one ends up with $E(k) \approx E_0 - (\text{real constant})k_i^2$, which is real. Thus, the allowed complex wave vectors form a line that starts in the extremum point and points perpendicular to the real axis. The complex band structure of a solid is then the ensemble of real-valued energy eigenvalues on the domain of complex wave vectors.

A simple model system for an electronic dispersion relation in a periodic potential is the nearly-free-electron model [177], which exposes the electron to a weak periodic potential

$$V(r) = V_0 + V_g \cos(gr) \tag{5.2}$$

leading to the formation of a band gap. The complex band structure $E(k_r + i\,k_i)$ of such a model is shown in figure 5.1(a). The band gap is located at the border of the Brillouin zone, which is displayed at the center of the (real) k_r axis. For energies in the band gap,

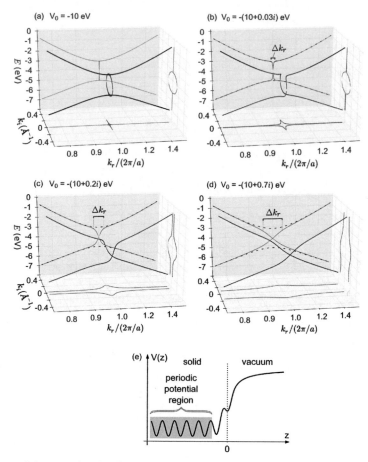

Figure 5.1: complex band structure for the one-dimensional nearly-free electron model [177] where the potential is given by $V(r)=V_0+V_g\cos(gr)$ and $g=3.5$Å^{-1} is the length of the shortest reciprocal lattice vector. The amplitude of the periodic component was chosen $V_g=1.0$ eV. Bloch waves with real wave vectors are drawn as black lines. (a) V_0 is real-valued; (b-d) V_0 attains an increasing imaginary part. The (k_r, E) plane is empha-sized with a darker shade. The dashed line reproduces the (k_r, E) projection of (a); Δk_r marks the range, where the (k_r, E) projection deviates more than a fixed threshold value from the respective curve in (a). (e) electronic potential in the vicinity of the surface. The gray shaded area marks the depth region, where the wave function may have the form given in equation (5.1) with complex wave vectors ($\Im(\vec{k})\neq 0$)

the wave vector becomes purely imaginary and forms a closed loop. Close to the band edges, the imaginary part is small, i.e. the wave function decays slowly spanning a depth of many atom layers, while for energies at the center of the band gap, the wave function decays more rapidly in space.

Photoemission probes all electronic states to which a plane wave coming from the vacuum can be coupled. This includes both categories: (1) Bloch waves extending across the crystal as well as (2) the exponentially decaying states ("evanescent states") described by inserting complex wave vectors in equation (5.1). The effect of inelastic scattering on the photoelectron (final state) is known to limit the escape depth and thereby the probing depth of photoemission. However, in strongly correlated materials, the rate of inelastic scattering processes may be considerably large for the initial states (e.g. the valence bands), as well. In a one-particle picture, these scattering processes can be mimicked by introducing an imaginary part of the potential, such that the potential acts as a sink for particles. According to Kevan [179], the Schroedinger equation for a quasiparticle reads (rewritten in SI units)

$$-\frac{\hbar^2 \nabla^2}{2m}\psi + V(\vec{r})\,\psi + \int d\vec{r}'\Sigma(\vec{r},\vec{r}',E)\psi(\vec{r}') = E\psi \qquad (5.3)$$

The imaginary part of the constant potential V_0 that we introduce, could thus be reinterpreted as $\Im(\Sigma_0)$ in a special type of self energy $\Sigma(\vec{r},\vec{r}',E)=\Sigma_0\,\delta(\vec{r}-\vec{r}')$, i.e. two electrons interact only, if they are in the same place. The complex band structure for increasing imaginary potential is shown in figures 5.1(b-d). Even now, solutions of the Hamilton operator with *real* energy eigenvalues (stationary states) exist. However, the spatial extent of the associated wave functions in $\vec{k}$-space becomes limited for *all* electronic states (i.e. all wave vectors attain a finite imaginary part), not just those which were originally in the band gap. In momentum-resolved photoemission, we explicitly select the real parts of the parallel momentum components (translating to $\Re(k_{||,x})$ and $\Re(k_{||,y})$), while the imaginary parts $\Im(k_{||,x})$, $\Im(k_{||,y})$ as well as the complex $k_\perp$ follow implicitly by boundary conditions of the inverse LEED state at the border between surface barrier potential and periodic bulk potential region (figure 5.1(e)). Hence, if we take k_r, k_i to be real and imaginary parts of a parallel wave vector component, e.g. $k_{||,x}=k_{||,x,r} + ik_{||,x,i}$, the photoemission energy spectrum $I(k_{||,x,r},E)$ represents a projection of the complex band structure into the (k_r, E) plane. In figures 5.1(a-d), this projection is displayed, as well as the projections into the (k_i, E) and the (k_r, k_i) planes.

The evolution of the (k_r, E)-plane projection for step-wise increasing imaginary part of the potential is interesting for the interpretation of photoemission energy spectra of materials where the rate of inelastic scattering processes in the initial state energy region is strong: For vanishing $\Im(V_0)$ (refer to equation (5.2), we have two parabolic bands separated by a band gap and a vertical line connecting the maximum of the lower band to the minimum of the upper band. When introducing a weak constant negative imaginary part to V_0 (figure 5.1(b)), the band dispersion in the (k_r, E)-plane is altered mainly around the extremal points, opening a gap Δk_r along the real axis of k, while the dispersion further away from the extremal points remains almost unaffected (compare to the black dashed curve). Further increasing the constant negative imaginary part of V_0 (figure 5.1(c)), the Δk_r-gap at the extremal points increases and the formerly vertical lines connecting the band extrema across the band gap in figure (a) become diagonal, i.e. the slope dE/dk_r decreases. Further increasing $-\Im(V_0)$ to an order of magnitude comparable to V_g (the amplitude of the oscillatory component of the potential) dissolves the band gap character at the Brillouin zone boundary (figure (d)): (1) The slope dE/dk_r in the region of the former band gap is hardly distinguishable from the slope dE/dk_r further away from the former band gap. (2) The wave vectors of the complex band structure have considerably large imaginary parts everywhere, not just in a small environment of the former band gap. Hence, for increasing magnitude of $-\Im(V_0)$, the separation between the usual Bloch waves with real $\vec{k}$ and band gap solutions with decaying amplitude ($\Im(k)\neq0$) is gradually removed as all electronic states with real energy eigenvalue attain an imaginary part of the wave vector, which means, they decay rapidly below the surface. In that case, i.e. strong inelastic scattering, it is to be expected that photoemission probes evanescent "band gap" states as well as the infinitely extended Bloch waves.

In the experimental photoemission spectra, we find several examples of vertical resonances (varying energy and constant $k_\parallel$) which start at extremal points of the bulk bands. Figure 4.6(d) shows vertical bars of high photoemission intensity, that start at an energy $E_i=E_F-1.5\,\text{eV}$ and span down to energies lower than $E_i=E_F-4.0\,\text{eV}$, with $k_{\parallel,[110]}=\pm1.2\text{Å}^{-1}$. More interestingly, figures 4.6(f) and 4.8 show another example: a band centered around $k_\parallel=0$, in the range $E_F-0.7\,\text{eV}<E_i<E_F-0.4\,\text{eV}$ whose minimum may be thought to occur at $E_i=E_F-0.7\,\text{eV}$ by extrapolation. However, a "gap of intensity" occurs at the position, where we would expect the minimum. Vertical bars of maximum intensity extend from the minimum energy of this band down to an energy $E_i\approx E_F-3.0\,\text{eV}$. They just about enclose the $k_\parallel$ range, along which the "gap of intensity" extends. This characteristic is rather similar to the (k_r, E) planar projection in figure 5.1(c).

We conclude, that the rather uncommon spectral features – vertical bars of high intensity in $I(k_\parallel, E)$ spectra – are a manifestation of the complex band structure for the initial states of the photoemission process. From the calculated complex band structures for the model system of a nearly free electron with additional constant negative imaginary part of the potential $-\Im(V_0)$ we see a tendency, that the k-radius around band minima or maxima, where the parabolic dispersion is altered (Δk_r in figure 5.1), increases with the magnitude of $-\Im(V_0)$. Hence, the k-radius of the gap at band minima in momentum-resolved photoemission spectra may be considered as a measure of the rate of inelastic scattering processes and thereby the strength of the electron-electron interaction. At the same time, that this gap opens, the originally infinite slope dE/dk_r of the vertical high-intensity features becomes finite and decreases.

In the present literature, the effect of an imaginary component of the potential has been shown for a modified Kronig-Penney model [180]. The original Kronig-Penney model results in a series of bands separated by band gaps, and the introduction of an imaginary component of the potential influences the band dispersion just around the band gaps, and leads to the removal of the band gap in a way that, e.g. the lowest band does not reach the Brillouin zone border, but joins the next higher-energetic band at a finite k-distance to the Brillouin zone border.

The attempt to illustrate the effect of an imaginary potential on the complex band structure and interpret some our photoemission spectral features in this way has been inspired by a work by Samarin et al.[181]. There, calculated band structures of W(110) (in the unoccupied energy range) are compared for the cases $\Im(V)=0$ and $\Im(V)=-0.5\,\text{eV}$ (as well as for inclusion or neglect of spin-orbit interaction). The modification of quasiparticle bands near the extremal points of their dispersion by introducing $\Im(V)=-0.5\,\text{eV}$ proceeds in a similar way as in figure 5.1: The extremal points are removed and replaced by branches of complex wave vector bands which span a large energy range (>10 eV). Some of those bands do not move on the real k axis (where we used the term "vertical bars" for our $I(k_{||}, E)$ photoemission spectra), and one of the complex bands spans more than 10 eV at the Brillouin zone border.

5.2 Self-energy lifetime from photoemission linewidths

In this section, the imaginary part of the momentum- and energy-resolved self energy in the valence band region of cobalt is extracted by a lineshape analysis of photoemission energy spectra. We assume that peaks in the photoemission intensity can be modelled by Lorentz peaks. This requires that (1) the variation of the transition matrix element in equation (2.3) for a single photoemission peak can be neglected, (2) the convolution of initial and final state spectral density in equation (2.10) produces a Lorentz peak and (3) quasiparticle peaks in the spectral density (equation (2.4)) are sufficiently described using constant values for $\Re(\Sigma)$ and $\Im(\Sigma)$. For increasingly broad peaks in photoemission energy spectra, the latter assumption becomes very inaccurate. Alternatively, lineshapes could be modelled using a more general form of a quasiparticle spectral density, as in equation (2.4), possibly approximating the self-energy by a local low-order polynomial, to keep the number of fit parameters low.

As a theoretical calculation [78] shows, the self-energy broadening due to electron-electron scattering for cobalt may become as large as 3.75 eV (corresponding to a full width of 3.2 eV) in the binding energy range $-8\,\text{eV} < E_i < -5\,\text{eV}$. For such an order of energy broadening, in addition to the above-mentioned insufficient description by Lorentz peaks, individual quasiparticle bands may overlap strongly and it may be impossible to disentangle them. Here, we focus on a binding energy range where theory data predicts a broadening of up to 0.5 eV and the use of Lorentz peaks was sufficient to reach satisfactory fitting of the modelled lineshapes to the experimental energy spectra as shown in figure 5.2. The model consists of a superposition of Lorentz peaks (representing quasiparticle peaks), multiplied by a Fermi distribution and subsequently convolved with a Gaussian curve representing the instrumental broadening (i.e. the resolution of the energy analyzer).

Figure 5.3(a) shows the results of modelling photoemission energy spectra for a range of $k_{||}$ values in the (110) direction which were obtained by using p-polarized light with a photon energy $h\nu=50\,\text{eV}$. Each modelled energy spectrum contained up to 5 Lorentz peaks, of which those have been selected that yielded *continously* varying parameters as a function of $k_{||}$. The Lorentz peak parameters are (1) energetic position, (2) half width and (3) area. The three coordinate axes contain the first two of these parameters in addition to $k_{||}$. Thus, the dispersion of the quasi-particle band is reproduced in the $(E, k_{||})$ plane, and the dependence of W_E on energy and in-plane wave vector $k_{||}$ is displayed in two separate planes. Cyan and black dots belong to quasiparticle bands denoted as "P_1" and "P_2" in

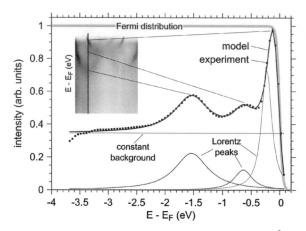

Figure 5.2: Modelling of an energy spectrum at selected $k_\| = -1.1\,\text{Å}^{-1}$ (full data in figure 4.6(d)) by superposed Lorentz peaks multiplied with Fermi statistics ($T = 300\,\text{K}$) and convoluted by a Gaussian distribution (full width at half maximum $w_G = 0.2\,\text{eV}$). Thus, the full widths of the Lorentz peaks obtained from the fitting of the model are clear of instrumental broadening effects.

figure 4.6(d), respectively. P_1 has been identified experimentally as a majority band, and P_2 as a minority band.

While we are interested in the electron-electron ("e-e") interaction as the primary broadening mechanism in cobalt, we need to consider that the resultant widths of the "deconvolved" Lorentz peaks still contain the influence of the final state lifetime. Secondly, the imaginary part of the self energy is composed of multiple inelastic scattering mechanisms.

$$\Im(\Sigma^{\text{tot}}) = \Im(\Sigma^{\text{e-e}}) + \Im(\Sigma^{\text{e-ph}}) + \Im(\Sigma^{\text{imp}}) \tag{5.4}$$

So far, electron-phonon coupling ("e-ph") and impurity scattering ("imp") have not been mentioned. In the following, these points are addressed before comparing the experimental results to theoretical data available in the literature:

According to section 2.4, the measured linewidths W_E have to be corrected to obtain the parameter $\Im(\Sigma_i)$. Solving equation (2.18) for $\Im(\Sigma_i)$ results in

$$\Im(\Sigma_i) = \frac{W_E}{2}\left(1 - \frac{v_i}{v_f}\right) + \Im(\Sigma_f)\frac{v_i}{v_f} \tag{5.5}$$

$v_f = dE_f/dk_\perp$ can be estimated from the free-electron dispersion (equation (2.5))

$$v_f\,[\text{eV Å}] \approx \pm 3.904\sqrt{E_f - E_{\text{vac}} + U_i\,(\text{eV})} \tag{5.6}$$

With $E_f - E_{\text{vac}} + U_i \approx h\nu + 12\,\text{eV} \approx 62\,\text{eV}$, we obtain $v_f \approx 30\,\text{eV Å}$. For the final state lifetime broadening we use data from [29] obtained for copper, with the full width of lifetime broadening amounting to $2\Im(\Sigma_f) = 6.0\,\text{eV}$ for electrons 50 eV above the Fermi level. Since

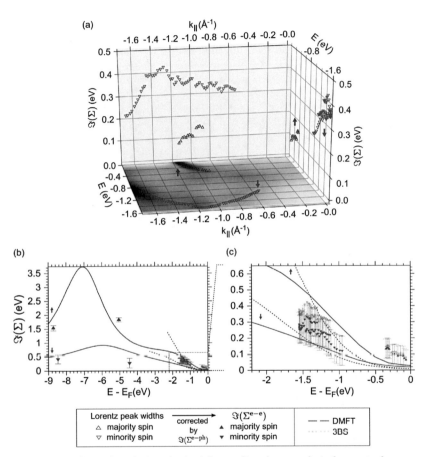

Figure 5.3: Lifetime broadening obtained from a line shape analysis for a set of energy spectra (differing in $k_{\parallel}$) taken from the data in figure 4.6(d); (a) The 3-dimensional plot shows the $E(k_{\parallel,[110]})$ dispersion and the half width of Lorentz peaks as a function of energy as well as a function of $k_{\parallel,[110]}$; (b,c) theoretical and experimental data for the imaginary part of the self energy as a function of binding energy. 3BS – 3-body scattering theory from [73]; DMFT – local-spin density approximation plus dynamical mean field theory from [78]; The error bars indicate the range of possible values of $\Im(\Sigma_i)$ after applying equation (5.5).

the dispersion of the initial states $E_i(k_\perp)$ in the surface normal direction is unknown, we estimate maximum values $|v_{i,\text{max}}^\downarrow| = 1.0\,\text{eV Å}$ in case of the minority-spin band, and $|v_{i,\text{max}}^\uparrow| = 0.65\,\text{eV Å}$ for the majority-spin band. The latter value has to be chosen smaller, since otherwise, negative values for the initial state linewidth would be obtained[1]. Sweeping v_i through the range $-|v_{i,\text{max}}| < v_i < |v_{i,\text{max}}|$, we obtain a range of $\Im(\Sigma_i)$ values from equation (5.5) which is indicated as errorbars in figure 5.3(b), where they are centered on the raw data (circles). The values then represent the total energetic broadening $\Im(\Sigma^{\text{tot}})$ as given in equation (5.4). In the following, we attempt to eliminate the electron-phonon scattering contribution. Since every value within the errorbars represents a possible value of $\Im(\Sigma^{\text{tot}})$, the errorbars in figure 5.3(c) have been shifted down by $\Im(\Sigma^{\text{e}-\text{ph}})$ and then represent a range of possible values for $\Im(\Sigma^{\text{e}-\text{e}}) + \Im(\Sigma^{\text{imp}})$, where $\Im(\Sigma^{\text{imp}})$ is constant and thus has to be smaller than the lowest occuring value of $\Im(\Sigma^{\text{tot}})$.

While the lifetime broadening due to electron-electron interaction vanishes at the Fermi level, the contribution from electron-phonon coupling does not for $T > 0\,\text{K}$. Within the Debye model and for temperatures $T \gg \theta_D$ (θ_D being the Debye temperature), the respective self energy term $\Im(\Sigma^{\text{e}-\text{ph}})$ becomes independent of the electron energy and can be written [182]

$$\Im(\Sigma^{\text{el}-\text{ph}}(T)) = \pi\,\lambda\,k_\text{B}\,T \tag{5.7}$$

where λ is called mass enhancement factor and is defined in terms of the Eliashberg function $\alpha^2\,F(E)$

$$\lambda = \int_0^{\hbar\omega_{\text{max}}^{\text{ph}}} \frac{\alpha^2 F(E)}{E}\,dE. \tag{5.8}$$

$F(E)$ is the density of phonon states, α is a coupling constant between phonons and electrons and $\omega_{\text{max}}^{\text{ph}}$ the highest phonon frequency. For hcp cobalt, $\lambda_\uparrow = 0.182$ and $\lambda_\downarrow = 1.134$ have been found by an ab initio calculation for majority and minority electrons, respectively [183]. Inserting $T = 300\,\text{K}$, we obtain $\Im(\Sigma_\uparrow^{\text{el}-\text{ph}}) = 0.015\,\text{eV}$ and $\Im(\Sigma_\downarrow^{\text{el}-\text{ph}}) = 0.092\,\text{eV}$. The contribution from electron-impurity scattering to the broadening $\Im(\Sigma^{\text{imp}})$ is a constant term that does not depend on energy or temperature. However, estimation of its size is difficult, as the relevant impurity concentration in the cobalt thin film is unknown and depends sensitively on the sample preparation[2].

Figure 5.3(c) contains the results of the previously discussed corrections to the experimental photoemission linewidths shown in figure 5.3(a). In case of the minority spin band, we find that after eliminating the broadening due to electron-phonon, the remaining $\Im(\Sigma^{\text{e}-\text{e}}) + \Im(\Sigma^{\text{imp}})$ is close to the theory for $\Im(\Sigma^{\text{e}-\text{e}})$ from [73]. The theoretical $\Im(\Sigma^{\text{e}-\text{e}})$ data for majority-spin quasiparticles close to the Fermi level ($E_\text{F} - 0.5\,\text{eV} < E_i < E_\text{F}$) does not exceed $0.05\,\text{eV}$. Here, we find a more significant deviation from theory. Since $\Im(\Sigma^{\text{e}-\text{e}}) = 0$ at the Fermi level, the data points close to the Fermi level give an upper bound for the

---

[1]Strictly, this argument requires that v_i does not change considerably along the $k_\|$ axis. This is likely the case, since $k_\perp(k_\|)$ is almost constant within the relevant $k_\|$ range (see figure 4.7). Furthermore $k_\perp$ is close to a central plane of the Brillouin zone, which is a mirror plane. v_i is zero on that plane and small in its vicinity. To have a measure of what "small" means, the order of v_i can be estimated from the dispersion of the majority-spin band in the *in-plane* direction: $|dE_i/dk_\|| < 0.5\,\text{eV Å}$ for $|k_\|| < 0.9\text{Å}^{-1}$.

[2]Experimental access is in principle possible by extrapolating temperature-dependent linewidths at the Fermi level to $T = 0\,\text{K}$. Under this condition, impurity scattering should be the only remaining broadening mechanism.

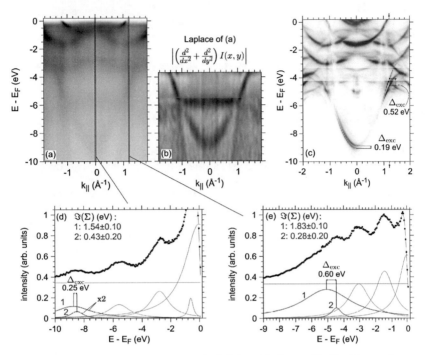

Figure 5.4: photoemission using p-polarized light with $h\nu = 50$ eV; 12 monolayers Co/Cu(001). (a) experimental photoemission intensity $I(k_{\|}, E)$ with $\vec{k}_{\|}$ in the [110] direction (b) Laplace operator (in arbitrary pixel size units) applied to the intensity in (a) emphasizes peak positions of the spectrum. (c) photoemission calculation with $\Im(\Sigma(E<E_F))=0.1$ eV (fully-relativistic layer KKR; omni [147]). (d,e) line shape analysis of two energy spectra from (a).

impurity scattering: $\Im(\Sigma^{\text{imp}}) < 0.05$ eV. The data points connected to the majority band show an increase $\Delta\Im(\Sigma^{\text{e-e}}) = 0.1$ eV across the energy range $E_F-0.4\,\text{eV}<E_i<E_F-0.05\,\text{eV}$, where the theory data increases by about 0.02 eV. Thus, experiment tells us that the majority band denoted as P_1 in figure 4.6 is subject to a sizeable electron-electron interaction already for energies close to the Fermi level, surpassing the theoretical prediction.

The calculated $\Im(\Sigma^{\text{e-e}})$ from [78] has a maximum at $E = E_F-7.1$ eV and decreases for more tightly bound electrons (see figure 5.3(b)), i.e. the lifetime of electrons increases towards the bottom of the valence band region and reaches values low enough such that quasiparticle resonances should be observable. It is interesting to check whether the same kind of decreasing energy broadening can be found in experiment. Figure 5.4(a) shows a photoemission spectrum covering a large binding energy range recorded with p-polarized radiation ($h\nu = 50$ eV). We find a weakly observable resonance (low peak to background intensity ratio) connected to the bottom-most valence band, which belongs to the highest symmetry representations ($\Delta_1, \Sigma_1, \Lambda_1$) of the respective symmetry groups (i.e. transforms like the scalar '1' under every operation of the wave-vector symmetry group). It appears more clearly in the contrast enhanced figure 5.4(b). A photoemission calculation representing the hypothetical, weakly-interacting variant of a 12 monolayer cobalt thin film on Cu(001) is presented in figure 5.4(c). The theoretical exchange splitting for the minimum of the valence band under consideration is reduced to $\Delta_{\mathrm{exc}} = 0.25\,\mathrm{eV}$ and increases significantly only for energies above $E_{\mathrm{F}} - 5\,\mathrm{eV}$ ($\Delta_{\mathrm{exc}} = 0.52\,\mathrm{eV}$). Hence, for the line shape model of a normal-emission photoemission spectrum, two Lorentz peaks with an initial-guess value of the energetic position of 8.0 eV and different widths have been used, to cover both majority and minority electron quasiparticles of the bottom-most valence band. All Lorentz peak parameters have been allowed to vary and the resulting model spectrum is shown in figure 5.4(d). The experimental exchange splitting $\Delta_{\mathrm{exc}}^{\mathrm{exp}} = 0.25\,\mathrm{eV}$ obtained in this way is close to the value $\Delta_{\mathrm{exc}}^{\mathrm{theo}} = 0.19\,\mathrm{eV}$ obtained by the fully-relativistic calculation. The same procedure has been performed for a photoemission spectrum at $k_{\parallel} = 1.2\,\mathrm{\AA}^{-1}$ (figure 5.4(e)) and confirms the agreement between experiment and theory with respect to Δ_{exc}. Furthermore, the higher-energetic peak "2" of the spin-pair has to be the minority-spin quasiparticle, and should as such exhibit a much smaller width, which was obtained by the fitting procedure without imposing any restriction on the peak positions and widths. The self energy imaginary parts corrected by the electron-phonon contribution $\Im(\Sigma^{\mathrm{e-ph}})$ are given in figures 5.4(d,e) with error ranges obtained by equation (5.5) using $|v_{i,\mathrm{max}}| = 2.0\,\mathrm{eV\,\AA}$. They have been added to figure 5.3(b), as well. We find, that $\Im(\Sigma)$ for the majority-spin quasiparticle is lower for $E_{\mathrm{F}} - 8.7\,\mathrm{eV}$ than for $E_{\mathrm{F}} - 5.0\,\mathrm{eV}$ in agreement with the non-monotonous theoretical curve 'DMFT' in figure 5.3(b). In case of minority quasiparticles, we find the opposite trend, and altogether, the values of $\Im(\Sigma)$ are drastically smaller than those for the majority spin in energy range $E_{\mathrm{F}} < 5.0\,\mathrm{eV}$.

5.3   Strong versus weak correlation

In the previous sections we discussed two observed characteristics in photoemission that are linked to the strong electron-electron interaction in cobalt: (1) The appearance of vertical bars (constant $k_{\parallel}$) of enhanced intensity in photoemission spectra $I(k_{\parallel}, E)$. (2) A strong spin-dependent increase of linewidths for decreasing initial state energies below the Fermi level.

Here, we compare photoemission spectra of fct cobalt thin films to those recorded on clean Cu(001). Both cobalt and copper are transition metals in the 4th row of the periodic table, with a rather small difference in the electronic configurations of the atoms (the copper atom has two additional electrons in the valence electronic shells). Based on this fact, the crystalline solids of these elements, which have comparable lattice constants, should have quite similar valence band structures (provided the same lattice type) with different positions of the Fermi level E_{F} (cf. band structures in an early calculation [184]). In case of copper, E_{F} is situated 2 eV above the d-bands, while for cobalt, it lies within the band width of the d-bands. The ferromagnetism of cobalt additionally shifts these bands (non-rigidly) to higher (minority spin) or lower (majority spin) energies. Spin-orbit coupling alters the symmetry groups of electronic wave functions and introduces some hybridization points. At this point, the band structures of fct cobalt and fcc copper would still allow to identify corresponding bands. The most striking difference arises due to the reduced lifetime of single-particle electronic states in cobalt with respect to those in copper, which is predominantly caused by the strong electron-electron interaction.

The comparison of photoemission spectra of clean fcc copper and fct Co/Cu(001) is one way to demonstrate the largely different degree of broadening in the valence bands between a strongly and a weakly correlated material. A second way to demonstrate the effect of strong correlation is to compare photoemission spectra for 12 monolayers Co/Cu(001) obtained by a fully-relativistic calculation (omni [147]) to the experimental data. The calculation has been performed using a constant self energy $\Sigma = (0 + 0.1i)$ eV for electron energies below the Fermi level (within the cobalt layer). This corresponds to hole lifetimes as found in weakly-correlated systems – for example, inverse lifetimes of d-band holes in Ag(001) have been found experimentally [185] to be below 0.3 eV for energies above $E_\mathrm{F} - 8.0$ eV.

Figures 5.5(b,e) summarize the spin-resolved photoemission measurements performed on Co/Cu(001) thin films at discrete binding energies, indicating the predominant spin character by upward (majority) and downward (minority) pointing triangles on top of a continous spin-integral $I(k_\parallel, E_i)$ spectrum. Thanks to the knowledge of the spin, experimental counterparts of the spin-split quasiparticle bands found in the calculated spectra in figures (a,d) can be established with more confidence. For example, the majority band marked with $P_1^\uparrow$ causes a high peak intensity near the Fermi level in figure 5.5(b). Its energetic minimum, which is clearly visible in the calculated spectrum (figure (a)) at $E_i = E_\mathrm{F} - 1.6$ eV, seems to be missing in the experimental spectrum. Nevertheless, due to the identified majority spin, it has to be the band denoted as $P_1^\uparrow$ in the calculated spectrum, since no other majority bands are present in this energy range and along the [110] direction.

A set of "landmarks" has been chosen, which are found both in the calculated and experimental spectra for Co/Cu(001) – mostly extremal points in the $E(k_\parallel)$ dependence formed by lines of maximum intensity. These landmarks do not occur at the same binding energy for figures (a,b) and (d,e). This can be attributed to the real part of the self energy: A quasiparticle peak in the experimental spectrum has its maximum intensity at $E_0(\vec{k}) + \Re(\Sigma(E, \vec{k}))$, while in the calculated spectrum, $\Re(\Sigma(E, \vec{k}))$ has been set to zero and the peak maximum occurs at $E_0(\vec{k})$. So we identify the difference, $\Delta E = E_\mathrm{exp} - E_\mathrm{theo}$ with the real part of the self energy. A plot of $\Delta E(k_\parallel, E_\mathrm{exp})$ versus E_exp is shown in figure 5.6, with annotations of the respective $k_\parallel$-values. The data indicate, that the $\vec{k}$ dependence of $\Re(\Sigma(E))$ is considerable. Several pairs of datapoints at approximately the same energy E_exp but different $k_\parallel$ or $k_\perp$ are present, that show a relatively large ΔE.

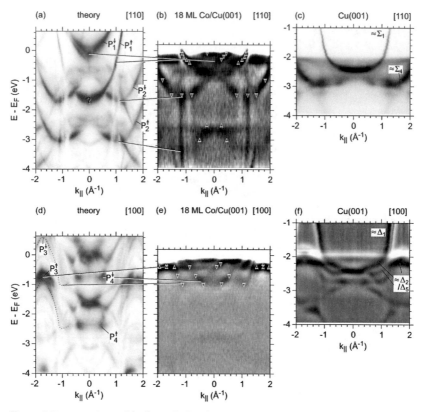

Figure 5.5: comparison of (a,d) a calculated spin-resolved photoemission spectrum for 12 ML Co/Cu(100) using $\Im(\Sigma) = 0.1\,\mathrm{eV}$; (b,c,e,f) contrast-enhanced experimental photoemission spectra (b,e) for 18 monolayers Co/Cu(001) and (c,f) for Cu(001); The spectra (c,f) have been shifted along the energy axis to place some of the corresponding photoemission resonances next to each other. The numbers at the energy axes give the true binding energies, in each case; Symmetry group representation symbols in (c,f) were assigned by a comparison to theoretical data in [186]. The up- and downward pointing triangles in (b,e) indicate the predominant spin polarization (majority and minority, respectively) obtained by spin-polarized photoemission measurements at discrete energies.

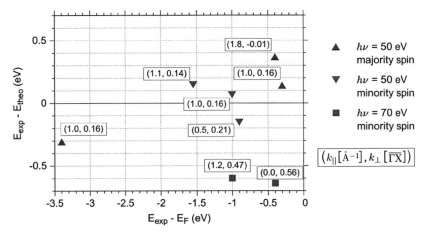

Figure 5.6: Energetic deviations of experimental photoemission resonances compared to the calculated photoemission spectra according to the correspondences shown in figure 5.5 and further spectra for $h\nu = 70\,\text{eV}$.

Chapter 6

Conclusion

This work presented a study of cobalt thin films on Cu(001) using a highly efficient spin- and momentum-resolved photoemission setup. As a prerequisite, an experimental procedure for spin-resolved measurements was developed, which includes (1) preparatory calibration measurements of the spin-sensitive electron mirror (the W(001) surface), (2) the spin-resolved measurement itself and (3) a non-trivial analysis which translates the recorded intensities to images of the spin polarization.

The Fermi surface of fcc cobalt has been recorded across an entire Brillouin zone scanning the photon energy in 1 eV steps from 35 eV to 200 eV and resolving at least 70×70 independent points per image. Interpretation in terms of a free-electron model for the photoemission final state has led to good agreement with calculated Fermi surfaces from [161]. Using an inner potential of 12.3 eV with respect to the Fermi level, the majority spin part yielded best agreement in an $I(k_{\|,x}, k_\perp)$ image, while the contours of the minority spin part were more easily identified in $I(k_{\|,x}, k_{\|,y})$ images for selected photon energies.

The valence band structure of fcc cobalt has been probed with spin-resolved photoemission using two different photon energies $h\nu$=50 eV, 70 eV, which cover the environment of the high-symmetry points Γ, X, W, K and L, as well as the high-symmetry axes Δ and Σ of the fcc Brillouin zone (see figure 7.4(a,b)). An experimental assignment of predominant majority and minority spin to the observed photoemission peaks has been summarized in figures 4.14(d,f,g) and 5.5(b,e). Quantitative information about the degree of spin-orbit coupling in the valence bands of cobalt has been obtained by changing the relative orientation of the magnetisation with respect to the incidence of the light and comparing the spin polarization values to those of the original geometry: asymmetries of the spin polarization up to 7% were found.

Quantum well states of Co thin films on Cu(001) have been investigated by two-photon photoemission (2PPE) for several thicknesses in the range from 3 to 13 monolayers. These are the quantum-confined equivalent of an unoccupied sp-band which touches the X point at $\approx E_F$+3.0 eV. Spin-resolved 2PPE measurement show a majority-spin peak originating from step-by-step transitions via the majority quantum well states (QWS), while no discernable enhancement of the minority photoelectron intensity is observed. The apparent absence of the minority QWS has been discussed in terms of the spin-dependent energetic broadening of electrons below and above the Fermi level as well as dipole selection rules.

With regard to the electron correlation, we have observed an uncommon type of spectral features – high-intensity patterns in $I(k_{\parallel,x}, k_{\parallel,y})$ spectra, which rigidly persist over a large range (several eV) on the energy axis and thus appear as vertical bars in $I(k_{\parallel}, E)$ spectra. We have discussed spatially decaying states with real eigenvalues and complex wave vectors that are a part of the complex band structure as a possible origin. The introduction of a constant imaginary part to the potential (acting as a sink for particles) emphasizes these states with respect to Bloch waves with real wave vectors. The "vertical bar" features are likely a manifestation of electron correlation for two reasons: (1) In energy spectra of clean Cu(001), we have not detected such spectral features. (2) Electron-electron correlation reduces the lifetime of single-electron states. The reduced lifetime is simulated by the imaginary part of the potential in a nearly-free electron model. This alters the dispersion relations of electronic bands around their extremal points in a way which resembles some of the quasiparticle peak dispersions found in $I(k_{\parallel}, E)$ spectra measured for cobalt.

Secondly, we have extracted linewidths which can be compared to the imaginary part of the self energy. For majority quasiparticles down to 0.5 eV below the Fermi level, we have found a deviation towards stronger broadening than in a published dataset calculated with dynamical mean-field theory (DMFT) [73, 78]. The more tightly bound minority quasiparticles, which we have analysed, fit closer to the same theory dataset.

For more extensive data regarding the self energy of electrons in cobalt, some improvements may be achieved in the future. The procedure to evaluate linewidths from photoemission spectra used in this work treats the dataset $I(k_{\parallel,x}, k_{\parallel,y}, E)$ of intensity versus wave vector and energy as independent energy spectra $I_{k_0}(E)$. For a particular photoemission peak which moves in energy as a function of the wave vector, this analysis generally produces discontinuities in the dispersion relation $E(k_{\parallel})$ as well as in the functional dependencies "linewidth versus energy" or "linewidth versus wave vector". A different approach should be used to truly exploit the fine grid of wave vectors and energies that are provided by momentum-resolved photoemission: Instead of choosing independent Lorentz peaks for each energy spectrum, every Lorentz peak parameter should be thought of as a function of the wave vector. These functions can be parameterized as polynomials, which enforces continuity on the previously mentioned functional dependencies of interest and – at the same time – accelerates the analysis while exploiting the full experimental data set. The motivation with regard to physics is to gain more experiment-based information on how the self energy depends on binding energy, wave vector, spin, orbital symmetries and possibly the localization of the electron, since these are questions of current theoretical interest, as well. The dependence on binding energy and spin was obtained for two quasiparticle bands. Some data points showed, that at the same energy, the imaginary part of the self energy was larger towards the border of the Brillouin zone. However, the picture is not very complete, yet. Furthermore, the considerable uncertainty due to the slope of the initial state bands needs to be removed, e.g. by an interpolated band structure which is required to evaluate for every wave vector in the Brillouin zone. Alternatively, photoemission calculations may be employed where the self energy parameters are adapted to fit the experimental data. Thereby the combination of initial and final state broadening and their respective slopes $dE/dk_{\perp}$ are treated correctly.

Future photoemission experiments may employ spin-sensitive electron mirrors with even higher efficiency and reliability. In this work, spin-resolved measurements were interrupted after two hours, to perform preparation procedures which recovered the clean W(100) surface and thus reverted the degradation of the efficiency. Recently, a pseudomorphic monolayer of Au on a (5×1) reconstructed Ir(001) surface has been proposed as a very promising candidate for multi-channel spin-polarimetry [187]. The structure can be reproducibly prepared and features spin asymmetries up to 77%, which are sustained in ultra-high vacuum for more than three weeks. Compared to the W(001) based imaging spin detector, the much longer lifetime saves the time of frequent preparation steps. Furthermore, the stability of the Au surface should strongly enhance the accuracy of the (delicate) calibration measurements, since these need to be done with the spin-analyzer surface matching as close as possible the condition (e.g. amount of adsorbates) during the actual measurement. As described in section 3.3.4, the precision of the calibration parameters $(S_{h,l}, R_{h,l})$ determine the error of the spin polarization. With the rapidly changing reflectivity and spin asymmetry of the W(100) surface, considerable effort was spent to minimize this kind of error. Therefore, a better stability of the spin-analyzer surface should considerably facilitate multi-channel parallelized spin-resolved photoemission experiments.

Appendix

7.1   2D color code for spin-polarized images

In order to represent spin polarised photoemission data over two-dimensional parameter spaces in an easily understandable way, we have used the following color code that represents both I_0 and P in one graph. When plotting a scalar function $z(x, y)$, it is customary to use a color code that translates z values to color values. Here, we need to represent two functions $P(k_{||,x}, k_{||,y})$ and $I_0(k_{||,x}, k_{||,y})$ in one color-encoded plot and thus we have to map pairs of (P, I_0) to a color. Therefore, we employ a two-dimensional color code. The mapping is based on the HSL (hue, saturation, lightness) color model. The sign of P determines the hue H: red for positive sign; blue for negative sign. The saturation S is proportional to the absolute value of P ($0 < |P| < 1$). The lightness L is proportional to the inverted, normalised spin-integral intensity $1 - I_0/I_{0,\text{max}}$. This is illustrated in figure 7.1. This color code has the following advantages: (1) Vanishing or very low photocurrent is always displayed as white (advantage over plotting P), regardless of the spin polarization; (2) unpolarized photocurrent can be distinguished from vanishing photocurrent as well as from spin-polarized photocurrent (advantage over plotting $P \times I_0$); (3) Spin-up and spin-down photocurrent can be identified by looking at a single graph (advantage over plotting $I_\uparrow$ and $I_\downarrow$ separately). Examples can be seen in figures 4.15 and 7.2.

7.2   Intensity asymmetries due to spin-orbit interaction

The asymmetry $A_{\text{s-o}}$ defined in equation (7.1) is related to spin-orbit coupling [12]. Instead of the spin-resolved partial intensities, it can be expressed by the average $(P_{\text{M+}} + P_{\text{M-}})/2$ of the spin polarization values for each polarity of the magnetization and the magnetic dichroism asymmetry A_{MD} defined in equation (7.4). The derivation starting from the

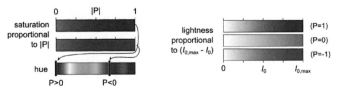

Figure 7.1: mapping of spin polarization P and spin-integral intensity I_0 to the HSL (hue, saturation, lightness) model

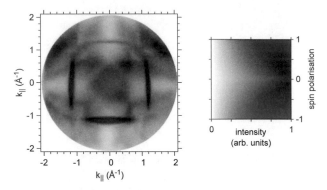

Figure 7.2: spin-resolved Fermi surface contour obtained by photoemission with p-polarized light and photon energy of 50 eV. Total (spin-integrated) intensity and spin polarization are mapped onto a 2-dimensional color code as given in the right-hand side figure. Thereby, the intensity shown in the momentum distribution is easily recognized as predominantly majority spin (red), minority spin (blue) or as unpolarized (gray).

definition of $A_{\text{s-o}}$ is briefly outlined here (index zero '0' means the sum of spin-up and spin-down):

$$A_{\text{s-o}} = \frac{I_{\uparrow,\text{M}+} - I_{\downarrow,\text{M}+} + I_{\uparrow,\text{M}-} - I_{\downarrow,\text{M}-}}{I_{\uparrow,\text{M}+} + I_{\downarrow,\text{M}+} + I_{\uparrow,\text{M}-} + I_{\downarrow,\text{M}-}} \tag{7.1}$$

$$A_{\text{s-o}} = \frac{P_{\text{M}+}\, I_{0,\text{M}+} + P_{\text{M}-}\, I_{0,\text{M}-}}{I_{0,\text{M}+} + I_{0,\text{M}-}} \tag{7.2}$$

$$A_{\text{s-o}} = (P_{\text{M}+} + P_{\text{M}-}) \frac{I_{0,\text{M}+}}{I_{0,\text{M}+} + I_{0,\text{M}-}} + P_{\text{M}-} \frac{I_{0,\text{M}-} - I_{0,\text{M}+}}{I_{0,\text{M}+} + I_{0,\text{M}-}} \tag{7.3}$$

$$A_{\text{MD}} = \frac{I_{0,\text{M}+} - I_{0,\text{M}-}}{I_{0,\text{M}+} + I_{0,\text{M}-}} \tag{7.4}$$

$$I_{0,\text{M}+} + I_{0,\text{M}-} = \frac{2 I_{0,\text{M}+}}{1 + A_{\text{MD}}} \tag{7.5}$$

$$A_{\text{s-o}} = \frac{P_{\text{M}+} + P_{\text{M}-}}{2} (1 + A_{\text{MD}}) - P_{\text{M}-} \cdot A_{\text{MD}} \tag{7.6}$$

$$A_{\text{s-o}} = \frac{P_{\text{M}+} + P_{\text{M}-}}{2} \quad \Leftrightarrow \quad (\text{if } A_{\text{MD}} = 0) \tag{7.7}$$

As the last line shows, $A_{\text{s-o}}$ is equivalent to the average $(P_{\text{M}+} + P_{\text{M}-})/2$, if the magnetic dichroism asymmetry A_{MD} vanishes. For evaluation from experimental data, $(P_{\text{M}+} + P_{\text{M}-})/2$ has the advantage that an experimental instability of the photoemission intensity between the measurements for 'M+' and 'M-' does not have any influence on $P_{\text{M}+}$ and $P_{\text{M}-}$, while it breaks the consistency of $I_{\uparrow,\text{M}+}, I_{\downarrow,\text{M}+}, I_{\uparrow,\text{M}-}, I_{\downarrow,\text{M}-}$ which are necessary to calculate $A_{\text{s-o}}$ from equation (7.1). The same procedure yields an identity for the asymmetry A_{ex} which is related to exchange interaction:

$$A_{\text{ex}} = \frac{I_{\uparrow,\text{M}+} - I_{\downarrow,\text{M}+} - I_{\uparrow,\text{M}-} + I_{\downarrow,\text{M}-}}{I_{\uparrow,\text{M}|} + I_{\downarrow,\text{M}|} + I_{\uparrow,\text{M}} + I_{\downarrow,\text{M}}} \tag{7.8}$$

$$A_{\text{ex}} = \frac{P_{\text{M}+} - P_{\text{M}-}}{2} \, (1 + A_{\text{MD}}) \, - \, P_{\text{M}} \cdot A_{\text{MD}} \tag{7.9}$$

7.3 Asymmetry relations for off-normal photoemission (cubic (001) surface)

For off-normal photoemission on a cubic (001) surface, we consider the geometry to be composed of the in-plane components of the photoelectron momentum $\vec{p}_{||} = \hbar \vec{k}_{||}$, the magnetization $\vec{M}$, the Poynting vector $\vec{S}_{||}$ (light propagation) and the y component of the spin polarization vector $\vec{P}$ is measured (i.e. y is the spin-sensitive axis). We assume that $\vec{M} = (0, M, 0)$; $\vec{S}_{||} = (0, S_{||,y})$ and that linearly polarized light (either s- or p-) is used. The

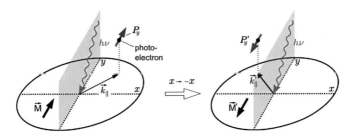

Figure 7.3: Effect of the mirror operation m_x ($x \to -x$) for a photoemission geometry where magnetization and azimuth of the incidence of light are parallel to the spin-sensitive axis (y). One of the mirror planes of the cubic (001) surface is aligned to the y-axis, as well.

effect of a mirror operation m_x ($x \to -x$) is shown in figure 7.3 (magnetization and spin polarization are axial vectors, which exhibit an additional sign flip when applying improper rotations, such as reflection). The azimuth of the incidence of light remains unaltered. In this particular geometry, the following identities for the photoemission intensity I and the spin polarization component P_y are obtained:

$$I(k_{||,x}, k_{||,y}, M, S_{||,y}) = I(-k_{||,x}, k_{||,y}, -M, S_{||,y}) \tag{7.10}$$

$$P_y(k_{||,x}, k_{||,y}, M, S_{||,y}) = -P_y(-k_{||,x}, k_{||,y}, -M, S_{||,y}) \tag{7.11}$$

The magnetic dichroism and the spin-orbit part $P_{\text{s-o}}$ of the spin polarization then have
the following properties:

$$A_{\text{MD}}(k_{||,x}, k_{||,y}) = \frac{I(k_{||,x}, k_{||,y}, M) - I(k_{||,x}, k_{||,y}, -M)}{I(k_{||,x}, k_{||,y}, M) + I(k_{||,x}, k_{||,y}, -M)} \tag{7.12}$$

$$A_{\text{MD}}(k_{||,x}, k_{||,y}) = \frac{I(-k_{||,x}, k_{||,y}, -M) - I(-k_{||,x}, k_{||,y}, M)}{I(-k_{||,x}, k_{||,y}, -M) + I(-k_{||,x}, k_{||,y}, M)} \tag{7.13}$$

$$A_{\text{MD}}(k_{||,x}, k_{||,y}) = -A_{\text{MD}}(-k_{||,x}, k_{||,y}) \tag{7.14}$$

$$P_{\text{s-o}}(k_{||,x}, k_{||,y}) = P_y(k_{||,x}, k_{||,y}, M) + P_y(k_{||,x}, k_{||,y}, -M) \tag{7.15}$$

$$P_{\text{s-o}}(k_{||,x}, k_{||,y}) = -P_y(-k_{||,x}, k_{||,y}, -M) - P_y(-k_{||,x}, k_{||,y}, M) \tag{7.16}$$

$$P_{\text{s-o}}(k_{||,x}, k_{||,y}) = -P_{\text{s-o}}(-k_{||,x}, k_{||,y}) \tag{7.17}$$

A_{MD} and $P_{\text{s-o}}$ vanish in normal emission ($k_{||} = 0$) and an integral of those quantities on a
field of view $\vec{k}_{||} < k_{||,\text{max}}$ vanishes as well.

7.4   Point groups in the fcc (and fct) lattice

Here, we illustrate symmetry properties of bulk electronic states in a face-centered cubic
crystal that are probed in photoemission measurements presented in sections 4.2.1 and
4.2.4. The Hermann-Mauguin notation is used (e.g. [89]): Numbers n denote rotational
axes with angles $\frac{360°}{n}$, a horizontal bar above a number ($\bar{n}$) means rotation and subsequent
inversion, m means mirror operation and a fraction symbol $\frac{n}{m}$ means that a mirror plane
exists of which the normal vector is parallel to the rotational axis. Several symbols in a row
belong to symmetry operations, where the defining vectors (either mirror plane normal
vectors or vector along the rotation axis) point in different directions.

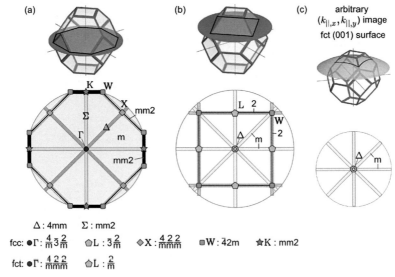

Figure 7.4: point groups of wave vectors $(k_{\|,x}, k_{\|,y}, k_\perp)$ in the fcc Brillouin zone for (a) $k_\perp=0$ (b) $k_\perp=\pi/c$ (c) arbitrary $k_\perp$; for the "fct" case (tetragonal distortion in the z-direction), the 3-fold rotational axes are absent, since these exchange the x, y, z directions, while those symmetry operations that transform $(x, y) \to (x'(x, y), y'(x, y))$ and $z \to z'(z)$, remain valid. (a) planar cut, whose environment is probed by $h\nu{=}50\,\mathrm{eV}$; (b) planar cut, whose environment is probed by $h\nu{=}70\,\mathrm{eV}$; (c) The minimum set of high-symmetry points for momentum-resolved images of photoemission on a fct (001) surface. These contain two inequivalent mirror planes and the Δ axis.

Bibliography

[1] M. M. H. Willekens, T. G. S. M. Rijks, H. J. M. Swagten, and W. J. M. de Jonge. Interface intermixing and magnetoresistance in Co/Cu spin valves with uncoupled Co layers. *J. Appl. Phys.* **78**, 7202 (1995).

[2] M. Guth, G. Schmerber, A. Dinia, D. Muller, and H. Errahmani. Giant magnetoresistance in Fe and Co based spin valve structures. *Phys. Lett. A* **279**, 255 (2001).

[3] S. S. P. Parkin, N. More, and K. P. Roche. Oscillations in exchange coupling and magnetoresistance in metallic superlattice structures: Co/Ru, Co/Cr, and Fe/Cr. *Phys. Rev. Lett.* **64**, 2304 (1990).

[4] I. McFadyen, E. Fullerton, and M. Carey. State-of-the-Art Magnetic Hard Disk Drives. *MRS Bull.* **31**, 379 (2006).

[5] G. A. Prinz. Magnetoelectronics. *Science* **282**, 1660 (1998).

[6] M. Johnson. Spin injection in metals and semiconductors. *Semicond. Sci. Technol.* **17**, 298 (2002).

[7] J. Islam, Y. Yamamoto, E. Shikoh, A. Fujiwara, and H. Hori. A comparative study of Co and Fe thin films deposited on GaAs(001) substrate. *J. Magn. Magn. Mater.* **320**, 571 (2008).

[8] C.-T. Chiang, A. Winkelmann, P. Yu, J. Kirschner, and J. Henk. Spin-orbit coupling in unoccupied quantum well states: Experiment and theory for Co/Cu(001). *Phys. Rev. B* **81**, 115130 (2010).

[9] M. Pickel, A. B. Schmidt, F. Giesen, J. Braun, J. Minár, H. Ebert, M. Donath, and M. Weinelt. Spin-Orbit Hybridization Points in the Face-Centered-Cubic Cobalt Band Structure. *Phys. Rev. Lett.* **101**, 66402 (2008).

[10] A. Fanelsa, E. Kisker, J. Henk, and R. Feder. Magnetic dichroism in valence-band photoemission from Co/Cu(001): Experiment and theory. *Phys. Rev. B* **54**, 2922 (1996).

[11] C. Schneider, M. Hammond, P. Schuster, A. Cebollada, R. Miranda, and J. Kirschner. Observation of magnetic circular dichroism in uv photoemission from ferromagnetic fcc cobalt films. *Phys. Rev. B* **44**, 12066 (1991).

[12] J. Henk. Temperature-dependent electronic structure, spin-resolved photoemission, and magnetic dichroism of ultrathin ferromagnetic films: Co/Cu(001). *J. Phys. Condens. Matter* **13**, 833 (2001).

[13] B. Kromker, M. Escher, D. Funnemann, D. Hartung, H. Engelhard, and J. Kirschner. Development of a momentum microscope for time resolved band structure imaging. *Rev. Sci. Instrum.* **79**, 53702 (2008).

[14] C. Tusche, M. Ellguth, A. A. Ünal, C.-T. Chiang, A. Winkelmann, A. Krasyuk, M. Hahn, G. Schönhense, and J. Kirschner. Spin resolved photoelectron microscopy using a two-dimensional spin-polarizing electron mirror. *Appl. Phys. Lett.* **99**, 032505 (2011).

[15] J. Pérez-Díaz and M. Muñoz. Fermi surface of ferromagnetic fcc cobalt. *Phys. Rev. B* **52**, 2471 (1995).

[16] J. Osterwalder, T. Greber, E. Wetli, J. Wider, and H. Neff. Full hemispherical photoelectron diffraction and Fermi surface mapping. *Prog. Surf. Sci.* **64**, 65 (2000).

[17] X. Gao, A. N. Koveshnikov, R. H. Madjoe, R. L. Stockbauer, and R. L. Kurtz. Dominance of the Final State in Photoemission Mapping of the Fermi Surface of Co/Cu(001). *Phys. Rev. Lett.* **90**, 37603 (2003).

[18] M. Mulazzi, J. Miyawaki, A. Chainani, Y. Takata, M. Taguchi, M. Oura, Y. Senba, H. Ohashi, and S. Shin. Fermi surface of Co(0001) and initial-state linewidths determined by soft x-ray angle-resolved photoemission spectroscopy. *Phys. Rev. B* **80**, 241106 (2009).

[19] D. Yu, M. Donath, J. Braun, and G. Rangelov. Spin-polarized unoccupied quantum-well states in ultrathin Co films on Cu(100). *Phys. Rev. B* **68**, 155415 (2003).

[20] C.-T. Chiang, A. Winkelmann, P. Yu, and J. Kirschner. Magnetic Dichroism from Optically Excited Quantum Well States. *Phys. Rev. Lett.* **103**, 77601 (2009).

[21] C.-T. Chiang, A. Winkelmann, J. Henk, F. Bisio, and J. Kirschner. Spin-selective pathways in linear and nonlinear photoemission from ferromagnets. *Phys. Rev. B* **85**, 165137 (2012).

[22] J. Stöhr and H. Siegmann. *Magnetism - From Fundamentals to Nanoscale Dynamics*. Springer Berlin / Heidelberg (2006). ISBN 978-3-540-30282-7.

[23] J. Hubbard. Electron Correlations in Narrow Energy Bands. *Proc. R. Soc. London, Ser. A* **276**, 238 (1963).

[24] H. Tasaki. Ferromagnetism in Hubbard Models. *Phys. Rev. Lett.* **75**, 4678 (1995).

[25] A. Vega and W. Nolting. Finite-temperature ferromagnetism of f.c.c. cobalt. *Phys. Status Solidi B* **193**, 177 (1996).

[26] G. Kotliar, S. Y. Savrasov, K. Haule, V. S. Oudovenko, O. Parcollet, and C. A. Marianetti. Electronic structure calculations with dynamical mean-field theory. *Rev. Mod. Phys.* **78**, 865 (2006).

[27] C. Calandra and F. Manghi. Three-body scattering theory of correlated hole and electron states. *Phys. Rev. B* **50**, 2061 (1994).

[28] S. Henning and W. Nolting. Ferromagnetism and nonlocal correlations in the Hubbard model. *Phys. Rev. B* **85**, 125114 (2012).

[29] J. Knapp, F. Himpsel, and D. Eastman. Experimental energy band dispersions and lifetimes for valence and conduction bands of copper using angle-resolved photoemission. *Phys. Rev. B* **19**, 4952 (1979).

[30] C. Schneider, P. Schuster, M. Hammond, H. Ebert, J. Noffke, and J. Kirschner. Spin-resolved electronic bands of FCT cobalt. *J. Phys. Condens. Matter* **3**, 4349 (1991).

[31] G. Margaritondo, D. L. Huber, and C. G. Olson. Photoemission spectroscopy of the high-temperature superconductivity gap. *Science* **246**, 770 (1989).

[32] D. Reznik, G. Sangiovanni, O. Gunnarsson, and T. P. Devereaux. Photoemission kinks and phonons in cuprates. *Nature* **455**, E6 (2008).

[33] S. Alvarado, H. Riechert, and N. Christensen. Spontaneous Spin Polarization of Photoelectrons from GaAs. *Phys. Rev. Lett.* **55**, 2716 (1985).

[34] W. Grobman and D. Eastman. Photoemission Valence-Band Densities of States for Si, Ge, and GaAs Using Synchrotron Radiation. *Phys. Rev. Lett.* **29**, 1508 (1972).

[35] F. Himpsel. Angle-resolved measurements of the photoemission of electrons in the study of solids. *Adv. Phys.* **32**, 1 (1983).

[36] A. Bostwick, J. McChesney, K. Emtsev, T. Seyller, K. Horn, S. Kevan, and E. Rotenberg. Quasiparticle Transformation during a Metal-Insulator Transition in Graphene. *Phys. Rev. Lett.* **103**, 056404 (2009).

[37] J. Nayak, M. Maniraj, A. Rai, S. Singh, P. Rajput, a. Gloskovskii, J. Zegenhagen, D. L. Schlagel, T. a. Lograsso, K. Horn, and S. R. Barman. Bulk Electronic Structure of Quasicrystals. *Phys. Rev. Lett.* **109**, 216403 (2012).

[38] M. P. Seah and W. A. Dench. Quantitative electron spectroscopy of surfaces: A standard data base for electron inelastic mean free paths in solids. *Surf. Interface Anal.* **1**, 2 (1979).

[39] R. Matzdorf. Investigation of line shapes and line intensities by high-resolution UV-photoemission spectroscopy - Some case studies on noble-metal surfaces. *Surf. Sci. Rep.* **30**, 153 (1998).

[40] B. Feuerbacher and R. F. Willis. Photoemission and electron states at clean surfaces. *J. Phys. C* **9**, 169 (1976).

[41] S. Hüfner. *Photoelectron Spectroscopy.* Springer Series in Solid-State Sciences. Springer Berlin, Heidelberg, 1st edition (1995). ISBN 3-540-19108-9.

[42] J. Pendry. Theory of photoemission. *Surf. Sci.* **57**, 679 (1976).

[43] E. Dietz and F. Himpsel. Photoemission via bloch states and evanescent band gap states for Cu(110). *Solid State Commun.* **30**, 235 (1979).

[44] P. Jennings and S. Thurgate. The inner potential in LEED. *Surf. Sci. Lett.* **104**, L210 (1981).

[45] V. N. Strocov. Bandstructure Effects in VeryLEED. *Int. J. Mod. Phys. B* **09**, 1755 (1995).

[46] V. Strocov, H. Starnberg, and P. Nilsson. Excited-state bands of Cu determined by VLEED band fitting and their implications for photoemission. *Phys. Rev. B* **56**, 1717 (1997).

[47] K. Capelle. A bird's-eye view of density-functional theory. *Braz. J. Phys.* **36**, 1318 (2006).

[48] M. Brack. The physics of simple metal clusters: self-consistent jellium model and semiclassical approaches. *Rev. Mod. Phys.* **65**, 677 (1993).

[49] R. M. Dreizler and E. K. U. Gross. *Density functional theory : an approach to the quantum many-body problem.* Springer Berlin (1990). ISBN 3-540-51993-9.

[50] D. M. Ceperley. Ground State of the Electron Gas by a Stochastic Method. *Phys. Rev. Lett.* **45**, 566 (1980).

[51] Z.-X. Shen and D. Dessau. Electronic structure and photoemission studies of late transition-metal oxides - Mott insulators and high-temperature superconductors. *Phys. Rep.* **253**, 1 (1995).

[52] D. McWhan, A. Menth, J. Remeika, W. Brinkman, and T. Rice. Metal-insulator transitions in pure and doped V_2O_3. *Phys. Rev. B* **7**, 1920 (1973).

[53] E. Dagotto. Correlated electrons in high-temperature superconductors. *Rev. Mod. Phys.* **66**, 763 (1994).

[54] P. Coleman, C. Pépin, Q. Si, and R. Ramazashvili. How do Fermi liquids get heavy and die? *J. Phys. Condens. Matter* **13**, R723 (2001).

[55] E. Morosan, D. Natelson, A. H. Nevidomskyy, and Q. Si. Strongly correlated materials. *Adv. Mater.* **24**, 4896 (2012).

[56] J. Quinn and R. Ferrell. Electron Self-Energy Approach to Correlation in a Degenerate Electron Gas. *Phys. Rev.* **112**, 812 (1958).

[57] A. Fetter and J. D. Walecka. *Quantum theory of many-particle systems.* International series in pure and applied physics. McGraw Hill, New York (1971). ISBN 0-07-020653-8.

[58] A. K. Ghosh, S. Chakraborty, and A. Manna. Density Functional Approach for Metals. The Matheiss Prescription. *Phys. Status Solidi B* **118**, 373 (1983).

[59] G. Baym and C. Pethick. *Landau Fermi-liquid theory. Concepts and Applications.* John Wiley and Sons, Inc. (1991). ISBN 0-471-82418-6.

[60] L. D. Landau. *JETP (Sov. Phys.)* **3**, 920 (1957).

[61] L. D. Landau. *JETP (Sov. Phys.)* **5**, 101 (1957).

[62] J. K. Grepstad, B. J. Slagsvold, and I. Bartos. Band structure dependent damping in photoemission from copper. *J. Phys. F* **12**, 1679 (1982).

[63] N. Smith, P. Thiry, and Y. Petroff. Photoemission linewidths and quasiparticle lifetimes. *Phys. Rev. B* **47**, 15476 (1993).

[64] J. Sánchez-Royo, J. Avila, V. Pérez-Dieste, and M. Asensio. Photoelectron lifetime determination of Ag(111) films at the Fermi surface. *Surf. Sci.* **482-485**, 752 (2001).

[65] P. Echenique, J. Pitarke, E. Chulkov, and A. Rubio. Theory of inelastic lifetimes of low-energy electrons in metals. *Chem. Phys.* **251**, 1 (2000).

[66] R. Knorren, K. Bennemann, R. Burgermeister, and M. Aeschlimann. Dynamics of excited electrons in copper and ferromagnetic transition metals: Theory and experiment. *Phys. Rev. B* **61**, 9427 (2000).

[67] T. Hertel, E. Knoesel, M. Wolf, and G. Ertl. Ultrafast Electron Dynamics at Cu(111): Response of an Electron Gas to Optical Excitation. *Phys. Rev. Lett.* **76**, 535 (1996).

[68] D. Pines and P. Nozières. *The theory of quantum liquids.* Addison-Wesley (1966). ISBN 0-201-09429-0.

[69] J. B. Pendry and J. F. L. Hopkinson. Photoemission from transition metal surfaces. *J. Phys. F* **8**, 1009 (1978).

[70] G. Tréglia, F. Ducastelle, and D. Spanjaard. Effect of coulomb correlations on energy bands in ferromagnetic transition metals : Ni, Co and Fe. *J. Phys.* **43**, 341 (1982).

[71] F. Manghi, V. Bellini, and C. A. Rozzi. Spin dependent many-body effects in the photoemission of Co. *J. Electron. Spectrosc. Relat. Phenom.* **140**, 523 (2004).

[72] M. I. Katsnelson and A. I. Lichtenstein. LDA++ approach to the electronic structure of magnets: correlation effects in iron. *J. Phys. Condens. Matter* **11**, 1037 (1999).

[73] J. Sánchez-Barriga, J. Braun, J. Minár, I. Di Marco, A. Varykhalov, O. Rader, V. Boni, V. Bellini, F. Manghi, H. Ebert, M. Katsnelson, A. Lichtenstein, O. Eriksson, W. Eberhardt, H. Dürr, and J. Fink. Effects of spin-dependent quasiparticle renormalization in Fe, Co, and Ni photoemission spectra:An experimental and theoretical study. *Phys. Rev. B* **85**, 205109 (2012).

[74] M. Steiner, R. Albers, and L. Sham. Quasiparticle properties of Fe, Co, and Ni. *Phys. Rev. B* **45**, 13272 (1992).

[75] M. Higashiguchi, K. Shimada, K. Nishiura, X. Cui, H. Namatame, and M. Taniguchi. Energy band and spin-dependent many-body interactions in ferromagnetic Ni(110): A high-resolution angle-resolved photoemission study. *Phys. Rev. B* **72**, 214438 (2005).

[76] C. Calandra and F. Manghi. Quasiparticle band structure of Ni and NiSi2. *Phys. Rev. B* **45**, 5819 (1992).

[77] K. Held, I. Nekrasov, G. Keller, V. Eyert, N. Blümer, A. McMahan, R. T. Scalettar, T. Pruschke, V. I. Anisimov, and D. Vollhardt. Realistic investigations of correlated electron systems with LDA + DMFT. *Phys. Status Solidi B* **243**, 2599 (2006).

[78] A. Grechnev, I. Di Marco, M. I. Katsnelson, A. I. Lichtenstein, J. Wills, and O. Eriksson. Theory of bulk and surface quasiparticle spectra for Fe, Co, and Ni. *Phys. Rev. B* **76**, 35107 (2007).

[79] G. Kotliar and D. Vollhardt. Strongly Correlated Materials: Insights From Dynamical Mean-Field Theory. *Phys. Today* **57**, 53 (2004).

[80] T. Maier, M. Jarrell, T. Pruschke, and M. Hettler. Quantum cluster theories. *Rev. Mod. Phys.* **77**, 1027 (2005).

[81] T. Miyake, C. Martins, R. Sakuma, and F. Aryasetiawan. Effects of momentum-dependent self-energy in the electronic structure of correlated materials. *Phys. Rev. B* **87**, 115110 (2013).

[82] A. Rubtsov, M. Katsnelson, and A. Lichtenstein. Dual fermion approach to nonlocal correlations in the Hubbard model. *Phys. Rev. B* **77**, 033101 (2008).

[83] W. Pauli. Zur quantenmechanik des magnetischen elektrons. *Zeitschrift für Physik A Hadrons and Nuclei* **43**, 601 (1927). 10.1007/BF01397326.

[84] P. A. M. Dirac. The quantum theory of the electron. *Proc. R. Soc. London, Ser. A* **117**, pp. 610 (1928).

[85] P. A. M. Dirac. The quantum theory of the electron. part ii. *Proc. R. Soc. London, Ser. A* **118**, pp. 351 (1928).

[86] J. Kessler. *Polarized Electrons*. Springer Series on Atoms and Plasmas. Springer-Verlag Berlin Heidelberg (1985). ISBN 3-540-15736-0.

[87] A. Messiah. *Quantenmechanik*, volume 2. de Gruyter (1976). ISBN 3110126699.

[88] C. M. Schneider and J. Kirschner. Spin- and angle-resolved photoelectron spectroscopy from solid surfaces with circularly polarized light. *Crit. Rev. Solid State Mater. Sci.* **20**, 179 (1995).

[89] M. Dresselhaus, G. Dresselhaus, and G. Jorio. *Group Theory. Application To Physics of Condensed Matter*. Springer-Verlag Berlin Heidelberg (2008). ISBN 978-3-540-32897-1.

[90] A. Rampe, G. Güntherodt, D. Hartmann, J. Henk, T. Scheunemann, and R. Feder. Magnetic linear dichroism in valence-band photoemission : Experimental and theoretical study of Fe(110). *Phys. Rev. B* **57**, 14370 (1998).

[91] J. Henk, T. Scheunemann, S. V. Halilov, and R. Feder. Magnetic dichroism and electron spin polarization in photoemission: analytical results. *J. Phys. Condens. Matter* **8**, 47 (1996).

[92] M. Cardona, N. Christensen, and G. Fasol. Relativistic band structure and spin-orbit splitting of zinc-blende-type semiconductors. *Phys. Rev. B* **38**, 1806 (1988).

[93] Y. Koroteev, G. Bihlmayer, J. Gayone, E. Chulkov, S. Blügel, P. Echenique, and P. Hofmann. Strong Spin-Orbit Splitting on Bi Surfaces. *Phys. Rev. Lett.* **93**, 046403 (2004).

[94] W. Kuch and C. M. Schneider. Magnetic dichroism in valence band photoemission. *Rep. Prog. Phys.* **64**, 147 (2001).

[95] C. Kentsch, M. Kutschera, M. Weinelt, T. Fauster, and M. Rohlfing. Electronic structure of Si(100) surfaces studied by two-photon photoemission. *Phys. Rev. B* **65**, 035323 (2001).

[96] S. Pawlik, R. Burgermeister, M. Bauer, and M. Aeschlimann. Direct transition in the system Ag(111) studied by one- and two-photon photoemission. *Surf. Sci.* **402-404**, 556 (1998).

[97] K. Giesen, F. Hage, F. Himpsel, H. Riess, and W. Steinmann. Two-photon photoemission via image-potential states. *Phys. Rev. Lett.* **55**, 300 (1985).

[08] M. Weinelt, A. Schmidt, M. Pickel, and M. Donath. Spin-polarized image-potential-state electrons as ultrafast magnetic sensors in front of ferromagnetic surfaces. *Prog. Surf. Sci.* **82**, 388 (2007).

[99] S. Schuppler, N. Fischer, T. Fauster, and W. Steinmann. Bichromatic two-photon photoemission spectroscopy of image potential states on Ag(100). *Appl. Phys. A* **51**, 322 (1990).

[100] R. Fischer, N. Fischer, S. Schuppler, T. Fauster, and F. Himpsel. Image states on Co(0001) and Fe(110) probed by two-photon photoemission. *Phys. Rev. B* **46**, 9691 (1992).

[101] T. Wegehaupt, D. Rieger, and W. Steinmann. Observation of empty bulk states on Cu(100) by two-photon photoemission. *Phys. Rev. B* **37**, 10086 (1988).

[102] H. Petek, H. Nagano, and S. Ogawa. Hot-electron dynamics in copper revisited: The {d-band} effect. *Appl. Phys. B* **68**, 369 (1999).

[103] H. Petek, H. Nagano, and S. Ogawa. Hole Decoherence of d Bands in Copper. *Phys. Rev. Lett.* **83**, 832 (1999).

[104] H. Petek and S. Ogawa. Femtosecond time-resolved two-photon photoemission studies of electron dynamics in metals. *Prog. Surf. Sci.* **56**, 239 (1997).

[105] S. Link, H. Dürr, G. Bihlmayer, S. Blügel, W. Eberhardt, E. Chulkov, V. Silkin, and P. Echenique. Femtosecond electron dynamics of image-potential states on clean and oxygen-covered Pt(111). *Phys. Rev. B* **63**, 115420 (2001).

[106] U. Höfer. Time-resolved coherent spectroscopy of surface states. *Appl. Phys. B* **68**, 383 (1999).

[107] A. Goris, K. M. Döbrich, I. Panzer, A. B. Schmidt, M. Donath, and M. Weinelt. Role of Spin-Flip Exchange Scattering for Hot-Electron Lifetimes in Cobalt. *Phys. Rev. Lett.* **107**, 026601 (2011).

[108] M. Hengsberger, F. Baumberger, H. Neff, T. Greber, and J. Osterwalder. Photoemission momentum mapping and wave function analysis of surface and bulk states on flat Cu(111) and stepped Cu(443) surfaces: A two-photon photoemission study. *Phys. Rev. B* **77**, 1 (2008).

[109] O. Schmidt, M. Bauer, C. Wiemann, R. Porath, M. Scharte, O. Andreyev, G. Schönhense, and M. Aeschlimann. Time-resolved two photon photoemission electron microscopy. *Appl. Phys. B* **74**, 223 (2002).

[110] H. Ueba and B. Gumhalter. Theory of two-photon photoemission spectroscopy of surfaces. *Prog. Surf. Sci.* **82**, 193 (2007).

[111] H. Ueba and T. Mii. Theory of energy- and time-resolved two-photon photoemission from metal surfaces - influence of pulse duration and excitation condition. *Appl. Phys. A* **71**, 537 (2000).

[112] K. Boger, M. Roth, M. Weinelt, T. Fauster, and P.-G. Reinhard. Linewidths in energy-resolved two-photon photoemission spectroscopy. *Phys. Rev. B* **65**, 075104 (2002).

[113] E. Knoesel, A. Hotzel, and M. Wolf. Temperature dependence of surface state lifetimes, dephasing rates and binding energies on Cu(111) studied with time-resolved photoemission. *J. Electron. Spectrosc. Relat. Phenom.* **88-91**, 577 (1998).

[114] T. Miller, A. Samsavar, G. Franklin, and T. Chiang. Quantum-Well States in a Metallic System: Ag on Au(111). *Phys. Rev. Lett.* **61**, 1404 (1988).

[115] B. Segall. Fermi Surface and Energy Bands of Copper. *Phys. Rev.* **125**, 109 (1962).

[116] J. Ortega, F. Himpsel, G. Mankey, and R. Willis. Quantum-well states and magnetic coupling between ferromagnets through a noble-metal layer. *Phys. Rev. B* **47**, 1540 (1993).

[117] L. Aballe, C. Rogero, and K. Horn. Quantum-size effects in ultrathin Mg films: electronic structure and collective excitations. *Surf. Sci.* **518**, 141 (2002).

[118] A. Shikin, D. Vyalikh, G. Prudnikova, and V. Adamchuk. Phase accumulation model analysis of quantum well resonances formed in ultra-thin Ag, Au films on W(110). *Surf. Sci.* **487**, 135 (2001).

[119] M. Milun, P. Pervan, and D. P. Woodruff. Quantum well structures in thin metal films: simple model physics in reality? *Rep. Prog. Phys.* **65**, 99 (2002).

[120] N. Smith, N. Brookes, Y. Chang, and P. Johnson. Quantum-well and tight-binding analyses of spin-polarized photoemission from Ag/Fe(001) overlayers. *Phys. Rev. B* **49**, 332 (1994).

[121] P. Echenique and J. Pendry. Theory of image states at metal surfaces. *Prog. Surf. Sci.* **32**, 111 (1989).

[122] E. McRae and M. Kane. Calculations on the effect of the surface potential barrier in LEED. *Surf. Sci.* **108**, 435 (1981).

[123] T.-C. Chiang. Photoemission studies of quantum well states in thin films. *Surf. Sci. Rep.* **39**, 181 (2000).

[124] W. McMahon, T. Miller, and T.-C. Chiang. Avoided crossings of Au(111)/Ag/Au/Ag double-quantum-well states. *Phys. Rev. Lett.* **71**, 907 (1993).

[125] W. E. McMahon, T. Miller, and T.-C. Chiang. A theoretical and experimental study of electronic confinement, coupling, and translayer interaction in noble-metal quantum-well structures. *Mod. Phys. Lett. B* **08**, 1075 (1994).

[126] M. Mueller, T. Miller, and T.-C. Chiang. Determination of the bulk band structure of Ag in Ag/Cu(111) quantum-well systems. *Phys. Rev. B* **41**, 5214 (1990).

[127] D. Wegner, A. Bauer, and G. Kaindl. Electronic Structure and Dynamics of Quantum-Well States in Thin Yb Metal Films. *Phys. Rev. Lett.* **94**, 126804 (2005).

[128] A. Mugarza, A. Marini, T. Strasser, W. Schattke, A. Rubio, F. García de Abajo, J. Lobo, E. Michel, J. Kuntze, and J. Ortega. Accurate band mapping via photoemission from thin films. *Phys. Rev. B* **69**, 115422 (2004).

[129] M. Jałochowski. Experimental evidence for quantum size effects in ultrathin metallic films. *Prog. Surf. Sci.* **48**, 287 (1995).

[130] M. Jałochowski, M. Hoffman, and E. Bauer. Quantized Hall effect in ultrathin metallic films. *Phys. Rev. Lett.* **76**, 4227 (1996).

[131] B. Orr, H. Jaeger, and A. Goldman. Transition-Temperature Oscillations in Thin Superconducting Films. *Phys. Rev. Lett.* **53**, 2046 (1984).

[132] W. Weber, a. Bischof, R. Allenspach, C. Würsch, C. Back, and D. Pescia. Oscillatory magnetic anisotropy and quantum well states in Cu/Co/Cu(100) films. *Phys. Rev. Lett.* **76**, 3424 (1996).

[133] H. Wierenga, W. de Jong, M. Prins, T. Rasing, R. Vollmer, A. Kirilyuk, H. Schwabe, and J. Kirschner. Interface Magnetism and Possible Quantum Well Oscillations in Ultrathin Co/Cu Films Observed by Magnetization Induced Second Harmonic Generation. *Phys. Rev. Lett.* **74**, 1462 (1995).

[134] F. J. Himpsel, J. E. Ortega, G. J. Mankey, and R. F. Willis. Magnetic nanostructures. *Adv. Phys.* **47**, 511 (1998).

[135] M. Escher, N. Weber, M. Merkel, C. Ziethen, P. Bernhard, G. Schönhense, S. Schmidt, F. Forster, F. Reinert, B. Krömker, and D. Funnemann. Nanoelectron spectroscopy for chemical analysis: a novel energy filter for imaging x-ray photoemission spectroscopy. *J. Phys. Condens. Matter* **17**, S1329 (2005).

[136] N. F. Mott. The Scattering of Fast Electrons by Atomic Nuclei. *Proc. R. Soc. London, Ser. A* **124**, 425 (1929).

[137] H. Tolhoek. Electron Polarization, Theory and Experiment. *Rev. Mod. Phys.* **28**, 277 (1956).

[138] R. Raue, H. Hopster, and E. Kisker. High-resolution spectrometer for spin-polarized electron spectroscopies of ferromagnetic materials. *Rev. Sci. Instrum.* **55**, 383 (1984).

[139] R. Feder. Spin-polarised low-energy electron diffraction. *J. Phys. C* **14**, 2049 (1981).

[140] R. Bertacco, D. Onofrio, and F. Ciccacci. A novel electron spin-polarization detector with very large analyzing power. *Rev. Sci. Instrum.* **70**, 3572 (1999).

[141] A. Winkelmann, D. Hartung, H. Engelhard, C.-T. Chiang, and J. Kirschner. High efficiency electron spin polarization analyzer based on exchange scattering at FeW(001). *Rev. Sci. Instrum.* **79**, 83303 (2008).

[142] F. U. Hillebrecht, R. M. Jungblut, L. Wiebusch, C. Roth, H. B. Rose, D. Knabben, C. Bethke, N. B. Weber, S. Manderla, U. Rosowski, and E. Kisker. High-efficiency spin polarimetry by very-low-energy electron scattering from Fe(100) for spin-resolved photoemission. *Rev. Sci. Instrum.* **73**, 1229 (2002).

[143] J. Kirschner and R. Feder. Spin Polarization in Double Diffraction of Low-Energy Electrons from W(001): Experiment and Theory. *Phys. Rev. Lett.* **42**, 1008 (1979).

[144] C. Tusche, M. Ellguth, A. Krasyuk, A. Winkelmann, D. Kutnyakhov, P. Lushchyk, K. Medjanik, G. Schönhense, and J. Kirschner. Quantitative spin polarization analysis in photoelectron emission microscopy with an imaging spin filter. *Ultramicroscopy* **130**, 70 (2013).

[145] E. Tamura and R. Feder. Spin Polarization in Normal Photoemission by Linearly Polarized Light from Nonmagnetic (001) Surfaces. *Europhys. Lett.* **16**, 695 (1991).

[146] N. Irmer, R. David, B. Schmiedeskamp, and U. Heinzmann. Experimental verification of a spin effect in photoemission: Polarized electrons due to phase-shift differences in the normal emission from Pt(100) by unpolarized radiation. *Phys. Rev. B* **45**, 3849 (1992).

[147] J. Henk, H. Mirhosseini, P. Bose, K. Saha, N. Fomynikh, T. Scheunemann, S. V. Halilov, E. Tamura, and R. Feder. OMNI – Fully relativistic electron spectroscopy calculations (2008). The computer code is available from the authors. *Input files for photoemission calculations on copper (and Co thin films on Co/Cu(001)) have been kindly provided by Jürgen Henk. Parameters describing the light source, film thickness, self energy broadening, etc have been varied or adapted to our experiments by the author of this work.*

[148] G. H. Reiling. Characteristics of Mercury Vapor-Metallic Iodide Arc Lamps. *J. Opt. Soc. Am. A* **54**, 532 (1964).

[149] P. Anderson. The Work Function of Copper. *Phys. Rev.* **76**, 388 (1949).

[150] D. L. Fried. Noise in Photoemission Current. *Appl. Opt.* **4**, 79 (1965).

[151] K. Zakeri, T. Peixoto, Y. Zhang, J. Prokop, and J. Kirschner. On the preparation of clean tungsten single crystals. *Surf. Sci.* **604**, L1 (2010).

[152] H.-J. Herlt and E. Bauer. A very low energy electron reflection study of hydrogen adsorption on W(100) and W(110) surfaces. *Surf. Sci.* **175**, 336 (1986).

[153] R. Barker and P. Estrup. Hydrogen on Tungsten(100): Adsorbate-Induced Surface Reconstruction. *Phys. Rev. Lett.* **41**, 1307 (1978).

[154] K. Yonehara and L. Schmidt. A LEED study of structures produced by H_2 on W(100). *Surf. Sci.* **25**, 238 (1971).

[155] P. W. Tamm. Binding States of Hydrogen on Tungsten. *J. Chem. Phys.* **54**, 4775 (1971).

[156] K. Schubert. Ein Modell für die Kristallstrukturen der chemischen Elemente. *Acta Crystallogr., Sect. B* **30**, 193 (1974).

[157] O. Heckmann, H. Magnan, P. le Fevre, D. Chandesris, and J. Rehr. Crystallographic structure of cobalt films on Cu(001): elastic deformation to a tetragonal structure. *Surf. Sci.* **312**, 62 (1994).

[158] C. Schneider, P. Bressler, P. Schuster, J. Kirschner, J. de Miguel, and R. Miranda. Curie temperature of ultrathin films of fcc-cobalt epitaxially grown on atomically flat Cu(100) surfaces. *Phys. Rev. Lett.* **64**, 1059 (1990).

[159] J. Crangle. The magnetic moments of cobalt-copper alloys. *Philos. Mag.* **46**, 499 (1955).

[160] T. S. Choy, J. Naset, S. Hershfield, C. Stanton, and J. Chen. A database of Fermi surface in virtual reality modeling language (vrml). In *Archives of the Bulletin of the American Physical Society* (2000). presentation L36.042.

[161] T. S. Choy. The Fermi Surface Database, http://www.phys.ufl.edu/fermisurface/ (2007).

[162] J. Luttinger. Fermi Surface and Some Simple Equilibrium Properties of a System of Interacting Fermions. *Phys. Rev.* **119**, 1153 (1960).

[163] R. A. Ballinger and C. A. W. Marshall. Electronic structure of cobalt. *J. Phys. F* **3**, 735 (1973).

[164] J. Sánchez-Barriga, J. Minár, J. Braun, A. Varykhalov, V. Boni, I. Di Marco, O. Rader, V. Bellini, F. Manghi, H. Ebert, M. Katsnelson, A. Lichtenstein, O. Eriksson, W. Eberhardt, H. Dürr, and J. Fink. Quantitative determination of spin-dependent quasiparticle lifetimes and electronic correlations in hcp cobalt. *Phys. Rev. B* **82**, 104414 (2010).

[165] E. Kisker, W. Gudat, and K. Schröder. Observation of a high spin polarization of secondary electrons from single crystal Fe and Co. *Solid State Commun.* **44**, 591 (1982).

[166] J. Cornwell. *Group Theory in Physics - An introduction.* Academic Press (1997). ISBN 0-12-189800-8.

[167] F. Himpsel and D. Eastman. Experimental energy-band dispersions and magnetic exchange splitting for cobalt. *Phys. Rev. B* **21**, 3207 (1980).

[168] K. Miyamoto, K. Iori, K. Sakamoto, H. Narita, A. Kimura, M. Taniguchi, S. Qiao, K. Hasegawa, K. Shimada, H. Namatame, and S. Blügel. Spin polarized d surface resonance state of fcc Co/Cu(001). *New J. Phys.* **10**, 125032 (2008).

[169] A. B. Schmidt, M. Pickel, T. Allmers, M. Budke, J. Braun, M. Weinelt, and M. Donath. Surface electronic structure of fcc Co films: a combined spin-resolved one- and two-photon-photoemission study. *J. Phys. D* **41**, 164003 (2008).

[170] F. Bisio, M. Nývlt, J. Franta, H. Petek, and J. Kirschner. Mechanisms of High-Order Perturbative Photoemission from Cu(001). *Phys. Rev. Lett.* **96**, 087601 (2006).

[171] T. Allmers, M. Donath, J. Braun, J. Minár, and H. Ebert. d- and sp-like surface states on fcc Co(001) with distinct sensitivity to surface roughness. *Phys. Rev. B* **84**, 245426 (2011).

[172] H. Eckardt, L. Fritsche, and J. Noffke. Self-consistent relativistic band structure of the noble metals. *J. Phys. F* **14**, 97 (1984).

[173] Y. Wu, C. Won, E. Rotenberg, H. Zhao, F. Toyoma, N. Smith, and Z. Qiu. Dispersion of quantum well states in Cu/Co/Cu(001). *Phys. Rev. B* **66**, 245418 (2002).

[174] V. Petrov, F. Rotermund, F. Noack, J. Ringling, O. Kittelmann, and R. Komatsu. Frequency conversion of Ti:sapphire-based femtosecond laser systems to the 200-nm spectral region using nonlinear optical crystals. *IEEE J. Sel. Top. Quantum Electron.* **5**, 1532 (1999).

[175] M. Aeschlimann, M. Bauer, S. Pawlik, W. Weber, R. Burgermeister, D. Oberli, and H. Siegmann. Ultrafast Spin-Dependent Electron Dynamics in fcc Co. *Phys. Rev. Lett.* **79**, 5158 (1997).

[176] V. Zhukov, E. Chulkov, and P. Echenique. Lifetimes of Excited Electrons In Fe And Ni: First-Principles GW and the T-Matrix Theory. *Phys. Rev. Lett.* **93**, 096401 (2004).

[177] N. Memmel. Monitoring and modifying properties of metal surfaces by electronic surface states. *Surf. Sci. Rep.* **32**, 91 (1998).

[178] V. Heine. On the General Theory of Surface States and Scattering of Electrons in Solids. *Proc. Phys. Soc.* **81**, 300 (1963).

[179] S. D. Kevan, editor. *Angle-Resolved Photoemission.* Studies in Surface Science and Catalysis. Elsevier (1992). ISBN 0-444-88183-2.

[180] H. Jones. The energy spectrum of complex periodic potentials of the Kronig-Penney type. *Phys. Lett. A* **262**, 242 (1999).

[181] S. Samarin, J. Williams, A. Sergeant, O. Artamonov, H. Gollisch, and R. Feder. Spin-dependent reflection of very-low-energy electrons from W(110). *Phys. Rev. B* **76**, 125402 (2007).

[182] S. Hüfner. *Very High Resolution Photoelectron Spectroscopy.* Springer Berlin, Heidelberg, New York, 1st edition (2007). ISBN 3-540-68130-2.

[183] M. J. Verstraete. Ab initio calculation of spin-dependent electron-phonon coupling in iron and cobalt. *J. Phys. Condens. Matter* **25**, 136001 (2013).

[184] L. Mattheiss. Energy Bands for the Iron Transition Series. *Phys. Rev.* **134**, A970 (1964).

[185] A. Gerlach, K. Berge, T. Michalke, A. Goldmann, R. Müller, and C. Janowitz. High-resolution photoemission study of long-lived d-holes in Ag. *Surf. Sci.* **497**, 311 (2002).

[186] G. Burdick. Energy Band Structure of Copper. *Phys. Rev.* **129**, 138 (1963).

[187] J. Kirschner, F. Giebels, H. Gollisch, and R. Feder. Spin-polarized electron scattering from pseudomorphic Au on Ir(001). *Phys. Rev. B* **88**, 125419 (2013).

Publications

1. C. Tusche, M. Ellguth, A. A. Ünal, C.-T. Chiang, A. Winkelmann, A. Krasyuk, M. Hahn, G. Schönhense, and J. Kirschner. *Spin resolved photoelectron microscopy using a two-dimensional spin-polarizing electron mirror.* Applied Physics Letters, **99**, 032505 (2011)

2. C. Tusche, M. Ellguth, A. Krasyuk, A. Winkelmann, D. Kutnyakhov, P. Lushchyk, K. Medjanik, G. Schönhense, and J. Kirschner. *Quantitative spin polarization analysis in photoelectron emission microscopy with an imaging spin filter.* Ultramicroscopy, **130**, 70 (2013)

3. A. Winkelmann, C. Tusche, A. A. Ünal, M. Ellguth, J. Henk, J. Kirschner. *Analysis of the electronic structure of copper via two-dimensional photoelectron momentum distribution patterns.* New Journal of Physics **14**, 043009 (2012)

4. A. A. Ünal, A. Winkelmann, C. Tusche, F. Bisio, M. Ellguth, C.-T. Chiang, J. Henk, J. Kirschner. *Polarization dependence and surface sensitivity of linear and nonlinear photoemission from Bi/Cu(111).* Physical Review B **86**, 125447 (2012)

5. A. Winkelmann, A. A. Ünal, C. Tusche, M. Ellguth, C.-T. Chiang, J. Kirschner. *Direct k-space imaging of Mahan cones at clean and Bi-covered Cu(111) surfaces.* New Journal of Physics **14**, 083027 (2012)

6. A. Winkelmann, M. Ellguth, C. Tusche, A. A. Ünal, J. Henk, J. Kirschner. *Momentum-resolved photoelectron interference in crystal surface barrier scattering.* Physical Review B **86**, 085427 (2012)

7. M. Ellguth, M. Schmidt, R. Pickenhain, H. v. Wenckstern, M. Grundmann. *Characterization of point defects in ZnO thin films by optical deep level transient spectroscopy.* Phys. Status Solidi B **248**, 941 (2011)

8. M. Schmidt, M. Ellguth, R. Karsthof, H. v. Wenckstern, R. Pickenhain, M. Grundmann, G. Brauer F.C.C. Ling. *On the T2 trap in zinc oxide thin films.* Phys. Status Solidi B **249** 588 (2012)

9. M. Schmidt, K. Brachwitz, F. Schmidt, M. Ellguth, H. v. Wenckstern, R. Pickenhain, M. Grundmann. G. Brauer, W. Skorupa, *Nickel-related defects in ZnO - A deep-level transient spectroscopy and photo-capacitance study.* Phys. Status Solidi B **248** 1949 (2011)

10. M. Schmidt, M. Ellguth, T. Lüder, F. Schmidt, H. v. Wenckstern, R. Pickenhain, M. Grundmann, G. Brauer, W. Skorupa. *Defects in a nitrogen-implanted ZnO thin film.* Phys. Status Solidi B **247** 1220 (2010)

11. H. v. Wenckstern, K. Brachwitz, M. Schmidt, C. P. Dietrich, M. Ellguth, M. Stölzel, M. Lorenz, M. Grundmann. *The E3 defect in MgZnO.* J. Electr. Mat. **39** 584 (2010)

12. M. Schmidt, M. Ellguth, C. Czekalla, H. v. Wenckstern, R. Pickenhain, M. Grundmann, G. Brauer, W. Skorupa, M. Helm, Q. Gu, C. C. Ling. *Defects in zinc-implanted ZnO thin films.* J. Vac. Sci. Technol. B **27** 1597 (2009)

Curriculum vitae

Persönliche Daten

Name	Ellguth
Vorname	Martin
Geburtsdatum	21.03.1984
Geburtsort	Eilenburg
Staatsangehörigkeit	deutsch
Familienstand	ledig

Kontaktinformation

Email	mellguth@mpi-halle.mpg.de

Bildungsgang

10/2003-11/2008	Physikstudium an der Universität Leipzig Vertiefungsrichtung: Halbleiterphysik Diplomarbeit: *Untersuchung von Midgap-Zuständen im ZnO mittels optischer und kapazitätsspektroskopischer Methoden* Abschluss: Physik-Diplom
12/2009-12/2010	wissenschaftlicher Mitarbeiter an der Universität Leipzig
05/2010-11/2013	Doktorand am Max-Planck-Institut für Mikrostrukturphysik, Halle

Halle (Saale), 2015

Erklärung an Eides statt

Hiermit erkläre ich, dass ich die vorliegende Arbeit

A spin- and momentum-resolved photoemission study of strong electron correlation in Co/Cu(001)

selbstständig und ohne fremde Hilfe verfasst, andere als die von mir angegebenen Quellen und Hilfsmittel nicht benutzt und die den benutzten Werken wörtlich oder inhaltlich entnommenen Stellen als solche kenntlich gemacht habe.

Eine Anmeldung der Promotionsabsicht habe ich an keiner anderen Fakultät einer Universität oder Hochschule beantragt.

Halle (Saale), 2015

Martin Ellguth

Acknowledgements

I am grateful to Prof. Dr. Jürgen Kirschner for the opportunity to do my PhD in this exciting research field, for his continous support and scientific advice.

I am much indebted to Dr. Christian Tusche for his excellent mentoring, for sharing with me his extensive knowledge in physics, experimental techniques, computer-related techniques and his experience in presenting scientific results in various forms. His strong dedication to extending the frontiers of surface science and photoemission provided a fruitful environment for doing research and succeeding with my PhD project.

I thank Prof. Claus Michael Schneider and his coworkers for the very profitable collaboration that enabled several weeks of beam time at the synchrotron Elettra, Trieste, Italy. Dr. Christian Tusche, Dr. Alexander Krasyuk and Dr. Vitaliy Feyer deserve special mention for their heavy involvement in the time-consuming installation and the later deconstruction of the imaging spin detector at the NanoESCA beamline.

I thank PD Dr. Jürgen Henk for sharing his photoemission code and providing input files as well as instructions how to use it.

I enjoyed many discussions with Dr. Cheng-Tien Chiang, and his support getting to know the laser setup. I also thank him for the tasty Oolong which he brought as a gift from his holiday in Taiwan.

I also like to thank all colleagues which I have met at the institute for contributing to a nice working atmosphere, for nice discussions and for the opportunity of playing table tennis, a very welcome complement to the mental work.

I am very happy to have enjoyed the company and support of friends and, most importantly, my family.